T0215009

Lasers in
3D PRINTING AND
MANUFACTURING

World Scientific Series in 3D Printing

Series Editor: Chee Kai Chua *(Nanyang Technological University, Singapore)*

Published:

Vol. 1   Bioprinting: Principles and Applications
         by Chee Kai Chua and Wai Yee Yeong

Vol. 2   Lasers in 3D Printing and Manufacturing
         by Chee Kai Chua, Murukeshan Vadakke Matham and Young-Jin Kim

World Scientific Series in 3D Printing

Lasers in
3D PRINTING AND
MANUFACTURING

Chee Kai Chua
Murukeshan Vadakke Matham
Young-Jin Kim

NTU, Singapore

World Scientific

NEW JERSEY · LONDON · SINGAPORE · BEIJING · SHANGHAI · HONG KONG · TAIPEI · CHENNAI · TOKYO

Published by

World Scientific Publishing Co. Pte. Ltd.
5 Toh Tuck Link, Singapore 596224
USA office: 27 Warren Street, Suite 401-402, Hackensack, NJ 07601
UK office: 57 Shelton Street, Covent Garden, London WC2H 9HE

British Library Cataloguing-in-Publication Data
A catalogue record for this book is available from the British Library.

World Scientific Series in 3D Printing — Vol. 2
LASERS IN 3D PRINTING AND MANUFACTURING

ISBN 978-981-4656-41-2
ISBN 978-981-4656-42-9 (pbk)

Desk Editor: Amanda Yun

Printed in Singapore

FOREWORD

By Professor D.L. Bourell, The University of Texas, USA

Additive Manufacturing, also known as 3D Printing, has experienced an explosion of interest in the last 6 years. For many new to the field it is a bit startling to learn that the technology is almost 30 years old, with the first commercial additive manufacturing fabricator being delivered in 1987. The recent interest arises from multiple sources, including expiring founding patents which increased competition and government initiatives in advanced manufacturing which spotlight additive manufacturing. Media have covered the technology across the gamut: human interest stories, organ printing, home construction and food printing, to name a few.

The concept of building parts in an additive, layer-wise fashion dates back to the mid-1800s. In the period of 1968–1982, a number of additive manufacturing processes were invented, but they were not commercialised. This was largely due to the absence of distributed computing which became prominent in the early 1980s. Distributed computing is critical to additive manufacturing, as it captures the part's geometry, controls the fabricator and serves as the human–machine interface.

The vast majority of additive manufacturing fabricators are based on material extrusion. These popular, low-cost print machines have historically been used to print parts in reasonably accurate shapes but with properties substantially inferior to conventionally manufactured goods. The additive manufacturing processes covered in this book are associated with relatively expensive technologies that generate parts with relative accuracy and reasonable mechanical properties for engineering applications.

According to Terry Wohlers of Wohlers Associates, additive manufacturing has quadrupled in value worldwide over the last five years. It is anticipated that the industry will grow by a factor of five between 2015 and 2020. Accompanying this growth has been the development of standards writing committees such as the ISO TC261 and ASTM F42 Committees, who have provided a systematic framework for dealing with the critical aspects of additive manufacturing for industrial purposes.

The future of additive manufacturing is intriguing. Organ printing has received a lot of press coverage. While the technology is some years out, the potential of organ printing is staggering, impacting the entire organ transplanting infrastructure and health industries. Food printing will continue to grow and mature. Efforts are underway in research facilities and in the private sector to accelerate the rate at which additive manufactured parts are made. This is not only a convenience. It directly impacts part cost and therefore affects the utilisation of additive manufacturing in numerous market sectors. Part quality will improve over time, particularly for low-cost 3D printers.

Lasers are used commercially most popularly in two additive manufacturing processes: stereolithography and laser sintering/melting. The former uses a low-powered laser in the mW range to photolytically crosslink acrylic or epoxy thermoset polymers. Stereolithography is particularly attractive for parts where surface finish and transparency are desired. Laser sintering (polymers) and laser melting (metals) use varied types of lasers in the 20–250 W range to selectively melt pulverised feedstock in a powder bed. This process is very popular, particularly for structural service applications where part mechanical properties are important. In addition, lasers are used in other more niche applications in additive manufacturing such as the creation of extremely small parts in the 10–100s of micron size range.

Professor Chee Kai Chua is internationally known and is the leading AM researcher in Singapore. He has worked in the field for over 20 years. Currently, he is the Executive Director of the Singapore Centre for 3D Printing (SC3DP), housed within Nanyang Technological University, Singapore. The facilities and associated research are impressive, making a real contribution to the field, both inside and outside Singapore. I have known Professor Chua for almost 20 years. His research and contributions to the field of additive manufacturing are excellent. He is a world-recognised authority on the subject.

This book is an excellent compendium of information relating to the employment of lasers in additive manufacturing. The background information on laser optics, types of lasers and radiation safety will be useful particularly to readers who are new to the field and whose experience with lasers is limited. The coverage of manufacturing is broad, including both traditional additive manufacturing using lasers as well as related areas of machining, nanoscale patterning and ultra-short pulse manufacturing.

D.L. Bourell

Professor, The University of Texas, Austin, TX, U.S.A.

PREFACE

Technology and science have never failed to capture our fascination over the years and their advancement has obviously made our life easier. Change is something we all embrace and welcome into our lives, and science has paved the way for it. Additive Manufacturing (AM), popularly known as 3D printing, is one such technological advancement, which is playing an increasingly significant role in the manufacturing arena currently. AM has revolutionised how parts are made these days and how manufacturing can be carried out. With the advent of sophisticated and efficient laser systems, Laser-Assisted Manufacturing (LAM) and AM technologies have attracted significant interest over conventional machining methods. In fact, commercially available laser systems and 3D printing equipment have projected this to the top of the latest thrust of manufacturing technologies.

The major learning outcome of this book will be in the understanding of basics of lasers, optics and materials used for manufacturing and 3D printing. While the first part of the textbook focuses on the basics and insights to today's key laser-assisted manufacturing, the remaining chapters detail different laser-assisted 3D printing technologies and equipment. Materials in 3D printing and relevant manufacturing details are categorised in terms of material aspects into polymer, metal and ceramics. The textbook also gives a good account on the fundamentals of continuous wave (CW) and pulsed laser-assisted manufacturing with illustrative examples. Printing on semiconductor materials and photoresists become the target for a technological roadmap in meeting the forecasted technological nodes of sub-45nm. From this perspective, patterning and the printing of 2D and 3D structures on different materials are illustrated with fundamentals of interference and how they can be used for achieving the forecasted technological nodes. Finally, laser-assisted 3D printing and manufacturing generally employs high-power lasers; hence a complete understanding of the safety procedures in handling lasers and laser-based systems is very essential. A detailed account of this is given in the textbook and is expected to provide a better understanding of handling laser-based equipment with full safety to all end users.

In order for the graduate and undergraduate students in mechanical and precision engineering to practise the concepts and related contents in-depth, many problems with different perspectives have been included in this textbook.

This book will help tertiary lecturers, university professors and researchers in the related fields of LAM and 3D printing to practise and train, as well as help practitioners in the art to solve industry-relevant problems. We believe that the information and detailed illustrations presented in this book will also be of immense use to scientists and practitioners of AM technologies and LAM.

This book is expected to open newer pathways in both fundamental and applied research frontiers in the years to come.

Chua C. K.
Professor

Murukeshan V. M.
Associate Professor

KIM Y.-J.
Nanyang Assistant Professor

ACKNOWLEDGEMENTS

First, we would like to thank our respective spouses, Wendy, Sushama and Grimi, and our respective children, Cherie, Clement, Cavell Chua, Jishnu, Pranav and Taehee Kim for their patience, support and encouragement to complete this book. We are grateful to the administration of Nanyang Technological University (NTU) for valuable support, especially from the Singapore Centre for 3D Printing (SC3DP) and the School of Mechanical and Aerospace Engineering (MAE). In addition, would like to thank Lee Jian Yuan, Huang Sheng, Lee Hyub and Chun Byung Jae for their valuable contributions. VMM would also like to thank Chua Juen Kee and Sidharthan R who have contributed to interferometric lithography and Aswin H for the line drawings and support. We wish to express sincere appreciation to our special assistant Kum Yi Xuan for selfless help and immense effort in the coordination and timely publication of this book.

We would also like to extend our special appreciation to Professor David L. Bourell for his foreword.

The acknowledgements would not be complete without the contributions of the following companies for supplying and helping us with the information about their products they develop, manufacture or represent:

1. 3D Systems Inc., USA
2. EnvisionTEC., Germany
3. EOS GmbH, Germany
4. Nanoscribe GmbH, Germany
5. Optomec Inc., USA
6. SLM Solutions, Germany

Your suggestions, corrections and contributions will be appreciated and reflected on the later editions of this book.

Chua C. K.
Professor

Murukeshan V. M.
Associate Professor

KIM Y.-J.
Nanyang Assistant Professor

ABOUT THE AUTHORS

 **CHUA Chee Kai** is the Executive Director of the Singapore Centre for 3D Printing (SC3DP) and a full Professor of the School of Mechanical and Aerospace Engineering at Nanyang Technological University (NTU), Singapore. Over the last 25 years, Professor Chua has established a strong research group at NTU, pioneering and leading in computer-aided tissue engineering scaffold fabrication using various additive manufacturing techniques. He is internationally recognised for his significant contributions in bio-material analysis and rapid prototyping process modelling and control for tissue engineering. His work has since extended further into additive manufacturing of metals and ceramics for defence applications.

Professor Chua has published extensively with over 300 international journals and conferences, attracting close to 6000 citations, and has a Hirsch index of 37 in the Web of Science. His book, *3D Printing and Additive Manufacturing: Principles and Applications*, now in its fifth edition, is widely used in American, European and Asian universities and is acknowledged by international academics as one of the best textbooks in the field. He is the World No. 1 Author for the area of Additive Manufacturing and 3D Printing (or Rapid Prototyping as it was previously known) in the Web of Science, and is the most 'Highly Cited Scientist' in the world for that topic. He is the Co-Editor-in-Chief of the international journal, *Virtual & Physical Prototyping* and serves as an editorial board member of three other international journals. In 2015, he started a new journal, the *International Journal of Bioprinting* and is the current Chief Editor. As a dedicated educator who is passionate in training the next generation, Professor Chua is widely consulted on additive manufacturing (since 1990) and has conducted more than 60 professional development courses for the industry and academia in Singapore and the world. In 2013, he was awarded the "Academic Career Award" for his contributions to Additive Manufacturing (or 3D Printing) at the 6th International Conference on Advanced Research in Virtual and Rapid Prototyping (VRAP 2013), 1–5 October, 2013, at Leiria, Portugal.

Dr Chua can be contacted by email at mckchua@ntu.edu.sg.

 **Murukeshan Vadakke Matham** is an Associate Professor at the School of Mechanical and Aerospace Engineering, Nanyang Technological University (NTU), Singapore. He has extensive teaching experience of over 17 years in the area of optical engineering and laser machining. In addition, Dr Murukeshan has published heavily in the area of long, short and ultrashort pulse based laser material processing, micro and nanoscale patterning using conventional and near-field interference concepts. He has delivered over 50 keynote, plenary and invited talks at international workshops, conferences or forums and is the author of over 270 international articles, which include 157 international journal papers, more than 120 international conference proceedings papers or conferences and six book chapters. He has also published four papers in Nature Publication group journals as a lead corresponding author. Dr Murukeshan is also inventor or co-inventor of six awarded or filed patents and holds eight innovation disclosures. He serves in the editorial boards of international journals such as Nature publication group's journal, *Scientific Reports*; is joint editor of the *Journal of Holography and Speckle* (JHS) and the *International Journal of Optomechatronics* (IJO). He was also associate editor of the *Journal of Medical Imaging and Health Informatics* (JMIHI) until 2015. He also leads a research

group which focuses on cutting edge research on Laser-assisted Fabrication, Nanoscale Optics, Biomedical Optics and Applied Optics for Metrology. His research has fetched more than S$10 million through competitive and industry funding in the recent years. For his contributions, Dr Murukeshan has won many international recognitions and awards. He has supervised or co-supervised over 25 PhD students and many of his research students' papers have won awards at international conferences. He is currently the Deputy Director of the Centre for Optical and Laser Engineering (COLE), NTU, and a Fellow of the Institute of Physics.

Dr Murukeshan V M can be reached at mmurukeshan@ntu.edu.sg

KIM Young-Jin is a Nanyang Assistant Professor and NRF Fellow at the School of Mechanical and Aerospace Engineering, Nanyang Technological University (NTU), Singapore. For more than a decade, he has been actively involved in advancing ultrafast laser technologies over a broad spectral bandwidth and has applied them to high-impact metrological and manufacturing applications. He developed a series of crystal- and fibre-based frequency comb systems by stabilising to the atomic clock of a frequency standard; established the 'definition of a metre' by precisely measuring absolute distances and surface profiles using frequency combs; expanded the wavelength regime of the frequency comb from infrared to extreme ultraviolet regime by field enhancements; demonstrated the world's first femtosecond laser system in outer space and successfully operated the system over two years. He is the author of more than 175 publications, including peer-reviewed journal articles and conference proceedings. He also holds 21 patents under his name and his publications have been cited more than 1700 times. He currently has a H-index of 16.

Dr Kim Young-Jin can be reached at yj.kim@ntu.edu.sg.

LIST OF ABBREVIATIONS

2D	= Two-Dimensional
3D	= Three-Dimensional
3DP	= Three-Dimensional Printing
ABS	= Acrylonitrile Butadiene Styrene
AEL	= Accessible Emission Level
AM	= Additive Manufacturing
AOM	= Acousto Optic Modulator
BBO	= beta-BaB_2O_4
CAD	= Computer-Aided Design
CAGR	= Compound Annual Growth Rate
CCD	= Charge-Coupled Device
CMOS	= Complementary Metal-Oxide Semiconductor
CNC	= Computer Numerical Control
CNT	= Carbon Nano-Tube
CW	= Continuous-Wave
DIW	= Direct Ink Writing
DMD	= Direct Metal Deposition; Digital Mirror Device
DMLS	= Direct Metal Laser Sintering
DLP	= Digital Light Processing
DUV	= Deep Ultra-Violet
EB	= Electron Beam
EBM	= Electron Beam Melting
EM	= Electro Magnetic
FDM	= Fused Deposition Molding
FEA	= Finite Element Analysis
FEM	= Finite Element Method
FESEM	= Field Emission Scanning Electron Microscopy
FFF	= Fused Filament Fabrication

FWHM	= Full Width Half Maximum
GTT	= Glass Transition Temperature
GO	= Graphene Oxide
HAZ	= Heat Affected Zone
HG	= Harmonic Generation
HHG	= High Harmonic Generation
HWP	= Half-Wave Plate
IEC	= International Electrotechnical Commission
IL	= Interference Lithography
IOT	= Internet Of Things
IR	= Infra-Red
KTP	= $KTiOPO_4$
LENS	= Laser Engineered Net Shaping
LAM	= Laser-Assisted Manufacturing
LBM	= Laser Beam Machining
LBO	= LiB_3O_5
LOM	= Laminated Object Manufacturing
MEMS	= Micro-Electro-Mechanical Systems
MPE	= Maximum Permissible Exposure
MQC	= Material Quality Centre
NA	= Numerical Aperture
NC	= Numerical Control
NIR	= Near Infra-Red
NTM	= Non-Traditional Manufacturing
OFCS	= Optical Fibre Communication Systems
OD	= Optical Density
OPD	= Optical Path-length Difference
OLPC	= Online Laser Power Control
PCB	= Printed Circuit Board
PLA	= Polylactic Acid
PVA	= Polyvinyl Alcohol
QWP	= Quarter-Wave Plate
RGO	= Reduced Graphene Oxide
RP	= Rapid Prototyping
SDL	= Selective Deposition Lamination
SHG	= Second Harmonic Generation

SHS	= Selective Heat Sintering
SLA	= Stereolithography Apparatus
SLM	= Selective Laser Melting
SLS	= Selective Laser Sintering
TE	= Transverse Electric
THG	= Third Harmonic generation
TM	= Transverse Magnetic
TPP	= Two Photon Polymerization
UTS	= Ultimate Tensile Strength
UV	= Ultra-Violet

CONTENTS

Foreword   v
Preface   ix
Acknowledgements   xi
About the Authors   xiii
List of Abbreviations   xvii

Chapter 1   Introduction   1
    1.1   Lasers for Manufacturing   1
    1.2   Lasers for 3D Printing   2
    1.3   Current State of the Art   3
    1.4   The Book: Major Focus   7
             Problems   10
             References   10

Chapter 2   Lasers and Basic Optics for 3D Printing
             and Manufacturing   15
    2.1   Introduction   15
    2.2   Interaction of Light with Materials: Different
             Aspects   16
    2.3   Effect of Different Parameters   20
    2.4   Laser: Basic Components   24
    2.5   Laser Beam: Characteristics   26
    2.6   Laser Beam: Beam Forming Optics   30
    2.7   Representative Lasers in 3D Printing and
             Manufacturing   36
    2.8   Summary   43
             Problems   43
             References   44

Chapter 3   **Materials for Laser-based 3D Printing**
                and Manufacturing   **49**
    3.1   Introduction   49
    3.2   Materials for 3D Printing and Manufacturing   51
    3.3   Polymers   55
    3.4   Metals   60
    3.5   Ceramics   66
    3.6   Composites   69
    3.7   Other materials   70
    3.8   Optical Properties of Materials in Laser-based
                Processing   72
                Problems   87
                References   88

Chapter 4   **One-, Two-, and Three-Dimensional Laser-Assisted**
                Manufacturing   **93**
    4.1   Introduction   93
    4.2   Laser Beam Machining (LBM)   94
    4.3   Oxy-Laser Cutting   97
    4.4   Three-Dimensional Processes   99
    4.5   Basic Components of a LM System   100
    4.6   Effect of State of Polarisation (SOP) of Laser Beam
                on Machining   102
    4.7   Laser Equipment Characteristics   104
    4.8   CW Laser and UV Laser Machining   111
    4.9   UV Laser Machining: Salient Features   112
    4.10   Laser Machining Applications   116
    4.11   Summary   116
                Problems   117
                References   118

Chapter 5   **Laser-based 3D Printing**   **121**
    5.1   Introduction   121
    5.2   Liquid-Based 3D Printing   123
    5.3   Powder-Based 3D Printing   143
    5.4   Solid-Based 3D Printing   162
                Problems   166
                References   167

Chapter 6	**Advanced 3D Manufacturing: Micro and Nanoscale Patterning**	**175**
6.1	Introduction	175
6.2	Interference	176
6.3	Two-Beam Interference	178
6.4	Laser Interference Lithography	179
6.5.	Four-Beam Interference	183
6.6	Fabrication System and Methodology	184
6.7	Ultrafast Laser Machining	184
6.8	Summary	193
	Problems	193
	References	194

Chapter 7	**Laser Safety and Hazards**	**197**
7.1	Introduction	197
7.2	Basic Hazards from Lasers	198
7.3	Laser Ratings and Safety Standards	213
7.4	Radiation Exposure and Laser Goggle Selection	221
7.5	Precautions and Exemplary Hazardous Cases	229
	Problems	232
	References	233

Chapter 8	**Future Prospects**	**237**
8.1	Introduction	237
8.2	Trends in Next-Generation 3D Printing	238
8.3	Future Prospects for Each Technical Chapter	244
8.4	Future Applications of Laser-based 3D Printing	248
8.5	Summary	250
	Problems	251
	References	252

Chapter 1

INTRODUCTION

This book will cover the basics of lasers, optics and materials used for 3D printing and manufacturing. It will also include several case studies for readers to apply their understanding of the topics, provide sufficient theoretical background and insights to today's key laser-assisted AM processes, and conclude with the future prospects of this exciting technology. Non-Traditional Manufacturing (NTM) processes refer to processes in which non-traditional energy transfer mechanism and/or non-traditional media for energy transfer are involved. NTM processes can machine precision components from conventional and advanced materials such as superalloys, composites and ceramics. A reduced number of machining steps and a higher product quality are some of the advantages they offer. NTM offers an attractive alternative and often the only choice for the processing of these materials. The new and potential applications require enhanced process capabilities and more precise process design and optimisation capabilities. In response, substantial progress in research has been made in recent years, especially in the areas of process innovation, modelling, simulation, monitoring and control. Laser-assisted manufacturing and 3D printing falls into these categories. This book will give a clear and concise coverage of the fundamentals and the applied technological aspects of this exciting and highly impactful technology.

1.1 Lasers for Manufacturing

LASER is an acronym for Light Amplification by Stimulated Emission of Radiation. Lasers are naturally coherent and monochromatic sources of electromagnetic radiation with wavelengths in the bandwidth of ultraviolet (UV), to visible (VIS), to the infrared (IR). Depending on the type of laser and wavelength desired, the laser medium could be solid, liquid or gaseous.

They can deliver very high power or energy depending on whether they emit a continuous wave (CW) or pulses (short or ultrashort) [1–3]. These lasers can be focused to a precise spot of a certain size/dimension on a given substrate through any intervening medium [4,5]. This enables lasers to have wide-ranging applications in materials processing [6–10]. In 1960, Maiman [11] invented the ruby laser for which he was awarded the Nobel Prize. If we scan the history from then till now, the technology of lasers has advanced by leaps and bounds which have eventually led to the invention of many new lasers including semiconductor lasers, Nd:YAG lasers, CO_2 gas lasers, dye lasers, and the latest excimer lasers and femtosecond lasers. While Nd:YAG lasers and CO_2 gas lasers remove materials via physical mechanisms, short pulse lasers such as excimer and ultrashort pulse lasers are found to remove materials by way of ablation. The actual development of these industrial lasers began in the mid-1970s to meet different industry needs, such as cutting, welding, drilling and melting. During the 1980s and early 1990s, lasers were successfully applied for heating, cladding, alloying and even for controlled thin-film deposition. Since then, the technology has advanced and developed further, and has become an integral part of manufacturing macro, micro, meso and nanoscale parts and artefacts. This has served many industries ranging from semiconductors, to conventional manufacturing platforms, and the marine, aerospace and healthcare industries, to name a few.

1.2 Lasers for 3D Printing

As detailed in the previous section, it was evident that since the invention of lasers in the 1960s, the manufacturing sector has looked upon using this controllable optical energy for a variety of its processes. For the manufacturing sector, the existence of a vast scale of possible applications in combination with the growing market demand has led to lasers being used as an efficient technology driven rapid development in the industry over the last few decades.

One such technological advancement made possible by the advent of lasers is termed Additive Manufacturing (AM). Nowadays, AM, popularly known as 3D printing, plays an increasingly significant role in the manufacturing arena. AM has revolutionised how prototypes are made and how small batch manufacturing should be carried out [10,12–17]. It has helped in bringing up a much closer relation between digital CAD (Computer-Aided-Design) models and a physically complex part [18]. For an industry where time translates into money, AM provides ample opportunities for innovation.

Due to the high flexibility and high efficiency of lasers, laser-assisted manufacturing (LAM) and AM technologies are attracting much attention and interest over traditional methods [13,19–22]. This chapter mainly focuses on some of the basic facts and backgrounds of lasers and their applications in 3D printing and manufacturing. It also provides a brief account of the contents of each chapter in the book; to help the reader obtain a holistic learning experience.

1.3 Current State of the Art

Inventions of new laser sources and assisted technology have evolved by leaps and bounds during the past decade. Since the invention of lasers, industries have been able to utilise this controllable optical energy for various applications. Many industries today depend on such lasers for a variety of production line applications. In the case of the healthcare industry, especially in diagnostic and therapeutic applications, lasers have found tremendous applications. These include marking steel utensils and surgical assists when performing surgeries. Automotive and aerospace industries have also long benefitted from the advantages of the laser machining process; e.g. employing them to make car and aircraft parts.

It was found to be a paradigm shift in the making when laser technology was found to go hand in hand with 3D printing or rapid prototyping applications.

1.3.1 *Laser technology and 3D printing*

Though multiple methodologies and techniques for 3D printing systems exist, they generally follow the same approach described in the flowchart shown in Fig 1.1, below.

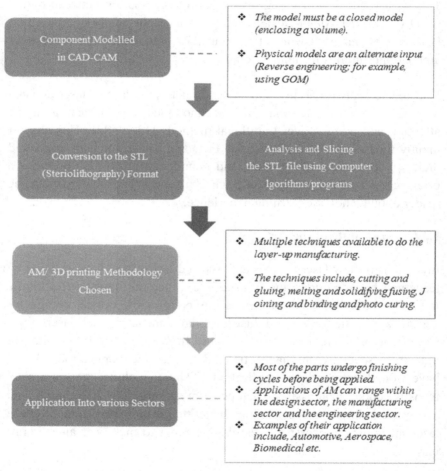

Fig. 1.1. Flowchart describing a general AM/3D printing process.

The flowchart describes the three important segments of a 3D printing process; namely, the required input (top), chosen methodology and materials (middle), and application (bottom). The printing process starts with designing a part or component using available CAD-CAM software, or by studying physical models using reverse engineering hardware (and software). The designed model is sliced by the computer software which determines the printing process. The second segment requires one to fix the process methodology and the material to be printed with. A multitude of methodologies are available for different types of materials. In this chapter, we focus mainly on the three most important methodologies in manufacturing that require the application of high-power lasers.

Selective Laser Sintering (SLS) is one of the methodologies to create 3D artefacts or features by employing a process called powder bed fusion. Generally, nylon is used as the material, and is transferred from bins containing fresh powder into the processing chamber using a recoating tool. This is followed by scanning the powder layers to sinter the particles together, eventually leading to the fabrication of a 3D layer of the artefact to be printed. Adjacent layers are done simultaneously; forming the solid part. It has to be mentioned that SLS doesn't need support structures as compared to conventional stereolithography or other similar techniques, considering the fact that the powder serves as a supporting material. This also enables the construction of more complicated geometric pieces.

There are a number of applications for 3D printing using SLS, such as the moving parts of advanced and conventional engines, architectural models, consumer products, electronics, auxiliary units and sculptures, and certain prosthetics used in medicine, to name a few [23–26]. Selective Laser Melting (SLM) is another methodology adopted in AM in order to print metal objects in 3D. In SLM, lasers do the softening of successive metallic powder layers, and the melting process is generally dictated by the CAD 3D files. This has found tremendous applications in the aerospace industry.

As the intricate parts can be printed using AM, these methods subdue the limitations associated with conventional manufacturing. Further, the manufacturing of metallic pieces with 3D printing can be done by using Direct Metal Laser Sintering (DMLS). The basic difference between the

SLM and DMLS is that SLM articulates the degree at which the powder particles are melted; whereas in DMLS, the melting is only done partially [27]. Furthermore, in DMLS, materials such as aluminium, steel, nickel, cobalt-chromium and titanium can also be used [28–30].

As far as the future of 3D printing is concerned, the number of consumer products and high-end technological users dependent on it is expected to grow by a huge amount in the coming years. Because of this perspective, more and more research and development in 3D printing has been done over the past 30 years. It can be summarised that, compared to conventional manufacturing, laser-assisted 3D printing can enable, with new synergy, the production of parts with complex configurations and accommodate design changes with high flexibility, to meet the rapidly changing landscape of various industries.

1.3.2 *Additive manufacturing: Advantages*

Additive Manufacturing (AM) promises the production of functional parts directly from their digital design. This allows for faster, and more accurate and reliable systems that can target small or large production capabilities, based on the type of system used and the complexity of the design. Using highly advanced AM systems available in the market, near perfect (compared to conventional machining) quality parts can be printed.

Production capabilities run hand in hand with technological developments. This demand arises from the requirement for a faster production capability for a lower cost and a low turnaround time. Technological advancements since the 1970s, such as Computer Numerical Control (CNC) and CAD-CAM to name a few, have helped in employing a faster, and more accurate and precise manufacturing process structure in the industries. To add on, AM has now provided the answer to better this process structure by employing automated, tool-less and pattern-less systems.

Being able to provide vast design freedom for the products manufactured, remain another notable advantage of the AM methodology. This allows product designers to worry less about the complexity of the parts being designed and focus more on targeting customer requirements. This in turn reduces the effect of part design on the lead time of the project, thus reducing the cost of the whole manufacturing cycle.

As described, AM being a tool-less manufacturing system reduces constraints on designing, manufacturing and verifying tooling systems. By reducing the overall part count, these tool-less systems can reduce the cost of the entire manufacturing and repair cycle of the machine.

By having a very low turnaround time, AM can cater variations in production capability for the present dynamic markets; meeting customers' needs. Also, AM allows a manufacturer to be updated on the technological front, gaining an economic advantage over their competition. With a few part count and flexible process control, AM offers added advantages in simplifying the supply chain and logistics sectors. All of these mentioned advantages of AM helps to shorten and simplify the manufacturing process promising a future where mass customisation prevails.

1.4 The Book—Major Focus

The major focus, summary of the table of contents, and learning objectives of the book are shown in Fig. 1.2. Different lasers and related basic optics that are essential for 3D printing and manufacturing are explained in detail. The fundamentals of light-matter interaction are discussed with illustrative examples, followed by the working principle of lasers, and their characteristics. All the essential basic optics and related optical components which form the basic optics for laser-assisted manufacturing and 3D printing are presented with illustrative examples. For realising higher resolution and higher productivity for laser-based 3D printing and manufacturing over various materials, photon interaction with target materials should be carefully considered. Materials in 3D printing and manufacturing are categorised in different ways depending on their applications; but generally, they can be classified as either polymer, metal or ceramic.

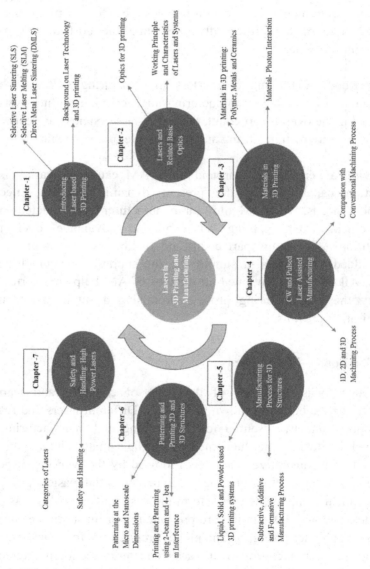

Fig. 1.2. Major contents of the text book to meet learning objectives.

The fundamentals of continuous wave (CW) and pulsed laser assisted manufacturing are discussed with illustrative examples in order to promote better understanding. An introduction on laser machining highlights the advantages and limitations compared to conventional machining processes in detail. How a single laser unit can be used to perform one-, two- and three-dimensional laser machining processes are discussed with illustrative examples.

The well-known fundamental manufacturing processes for 3D structures, such as the subtractive, additive and formative ones given in this book, will emphasise on how lasers can be applied to them as a highly efficient energy source. There are many ways in which one can classify the numerous 3D printing systems in the market, and the better ways to classify such 3D printing systems are explained by categorising them into (1) liquid-based, (2) solid-based and (3) powder-based systems. Generally speaking, patterning and printing 2D and 3D structures on different materials, which also include photosensitive materials, have in the recent past found potential scientific and industrial applications. The book also covers fundamental aspects of patterning at the micro and nanoscale dimensions using the principle of laser interference. Printing and patterning based on 2–beam and 4-beam interference with the help of a multiple beam interference lithography system are illustrated. This approach will be one of the many new methods coming up which are expected to change micro and nanoscale fabrication, as well as 3D printing at such smaller scales.

Laser-assisted 3D printing and manufacturing generally employs high power lasers whose power exceeds the certain thresholds required for the physical status change of materials. Hence, these high power lasers are very hazardous and a direct or an indirect exposure can burn the retina of the eye or the skin. So a complete understanding of the safety procedures for the handling of lasers is essential for any user. Future prospects of 3D printing and the role of lasers detailed in this book with representative examples are a good resource for prospective readers, and will help them envisage the growth of laser-based 3D printing and manufacturing and the very many impactful roles they play in the near future. Finally, this book is expected to be a resource book for researchers in their initial years. Hence the working principle of each individual process is described and a detailed update of the literature, scientific issues and technological innovations are covered by giving comprehensive state-of-the-art literature references at the end of each chapter.

Finally, this textbook is expected to become a timely information resource for undergraduates, postgraduates and researchers who are interested in this emerging technology.

Problems

1. How lasers remove materials during machining processes? Explain the difference between laser machining using CW lasers and pulsed lasers.

2. Explain briefly the steps involved in a general AM/3D printing process with a flowchart.

3. What are the advantages of AM being a tool-less manufacturing system?

4. How are 3D printing equipment and processes available in the market categorised?

5. Describe the three important segments of a 3D printing process. Correlate each segment with respective advantages and limitations.

References

[1]   H. Rong, R. Jones, A. Liu, O. Cohen, D. Hak, A. Fang *et al.*, "A continuous-wave Raman silicon laser," *Nature* **433**(7027), 725–728 (2005).

[2]   U. Keller, "Recent developments in compact ultrafast lasers," *Nature* **424**(6950), 831–838 (2003).

[3] P. Emma, R. Akre, J. Arthur, R. Bionta, C. Bostedt, J. Bozek *et al.*, "First lasing and operation of an ångstrom-wavelength free-electron laser," *nature photonics* **4**(9), 641-647 (2010).

[4] R.R. Gattass and E. Mazur, "Femtosecond laser micromachining in transparent materials," *Nature photonics* **2**(4), 219–225 (2008).

[5] A. Weck, T. Crawford, D. Wilkinson, H. Haugen and J. Preston, "Laser drilling of high aspect ratio holes in copper with femtosecond, picosecond and nanosecond pulses," *Applied Physics A* **90**(3), 537–543 (2008).

[6] C.B. Schaffer, A. Brodeur and E. Mazur, "Laser-induced breakdown and damage in bulk transparent materials induced by tightly focused femtosecond laser pulses," *Measurement Science and Technology* **12**(11), 1784 (2001).

[7] A. Zoubir, L. Shah, K. Richardson and M. Richardson, "Practical uses of femtosecond laser micro-materials processing," *Applied Physics A* **77**(2), 311-315 (2003).

[8] W. Steen, "Laser material processing—an overview," *Journal of Optics A: Pure and Applied Optics* **5**(4), S3 (2003).

[9] M. Malinauskas, A. Žukauskas, S. Hasegawa, Y. Hayasaki, V. Mizeikis, R. Buividas *et al.*, "Ultrafast laser processing of materials: from science to industry," *Light: Science & Applications* **5**(8), e16133 (2016).

[10] S. Perinchery, E. Smits, A. Sridhar, P. Albert, J. van den Brand, R. Mandamparambil *et al.*, "Investigation of the effects of LIFT printing with a KrF-excimer laser on thermally sensitive electrically conductive adhesives," *Laser Physics* **24**(6), 066101 (2014).

[11] T.H. Maiman, "Stimulated Optical Radiation in Ruby," *Nature* **187**(4736), 493–494 (1960).

[12] J. Loy and P. Tatham, "Redesigning Production Systems" In: *Handbook of Sustainability in Additive Manufacturing*, Springer (2016).

[13] R.M. Mahamood and E.T. Akinlabi, "Laser Additive Manufacturing," *Advanced Manufacturing Techniques Using Laser Material Processing*, 1 (2016).

[14] D. Lin, Q. Nian, B. Deng, S. Jin, Y. Hu, W. Wang *et al.*, "Three-Dimensional Printing of Complex Structures: Man Made or toward Nature?," *ACS Nano* **8**(10), 9710-9715 (2014).

[15]   T. Birtchnell and J. Urry, A New Industrial Future?: 3D Printing and the Reconfiguring of Production, Distribution, and Consumption, Routledge 2016.

[16]   C.K. Chua and K.F. Leong, 3D Printing and additive manufacturing: Principles and Applications. World Scientific, Singapore, (2014).

[17]   C. Chua, S. Chou and T. Wong, "A study of the state-of-the-art rapid prototyping technologies," *The International Journal of Advanced Manufacturing Technology* **14**(2), 146-152 (1998).

[18]   M. Naing, C. Chua, K. Leong and Y. Wang, "Fabrication of customised scaffolds using computer-aided design and rapid prototyping techniques," *Rapid Prototyping Journal* **11**(4), 249-259 (2005).

[19]   D.L. Bourell, "Perspectives on Additive Manufacturing," *Annual Review of Materials Research*(0), (2016).

[20]   D.I. Wimpenny, P.M. Pandey and L.J. Kumar, [Advances in 3D Printing & Additive Manufacturing Technologies] Springer, (2016).

[21]   C.K. Chua and W.Y. Yeong, Bioprinting: principles and applications, World Scientific (2014).

[22]   Z.X. Khoo, J.E.M. Teoh, Y. Liu, C.K. Chua, S. Yang, J. An *et al.*, "3D printing of smart materials: A review on recent progresses in 4D printing," *Virtual and Physical Prototyping* **10**(3), 103-122 (2015).

[23]   K. Tan, C. Chua, K. Leong, C. Cheah, P. Cheang, M.A. Bakar *et al.*, "Scaffold development using selective laser sintering of polyetheretherketone–hydroxyapatite biocomposite blends," *Biomaterials* **24**(18), 3115-3123 (2003).

[24]   C. Chua, K. Leong, K. Tan, F. Wiria and C. Cheah, "Development of tissue scaffolds using selective laser sintering of polyvinyl alcohol/hydroxyapatite biocomposite for craniofacial and joint defects," *Journal of Materials Science: Materials in Medicine* **15**(10), 1113-1121 (2004).

[25]   L.-E. Loh, C.-K. Chua, W.-Y. Yeong, J. Song, M. Mapar, S.-L. Sing *et al.*, "Numerical investigation and an effective modelling on the Selective Laser Melting (SLM) process with aluminium alloy 6061," *International Journal of Heat and Mass Transfer* **80**, 288-300 (2015).

[26]   K. Tan, C. Chua, K. Leong, C. Cheah, W. Gui, W. Tan *et al.*, "Selective laser sintering of biocompatible polymers for applications in tissue engineering," *Bio-medical materials and engineering* **15**(1, 2), 113-124 (2005).

[27]   M. Agarwala, D. Bourell, J. Beaman, H. Marcus and J. Barlow, "Direct selective laser sintering of metals," *Rapid Prototyping Journal* **1**(1), 26-36 (1995).

[28]   M. Khaing, J. Fuh and L. Lu, "Direct metal laser sintering for rapid tooling: processing and characterisation of EOS parts," *Journal of Materials Processing Technology* **113**(1), 269-272 (2001).

[29]   T. Grünberger and R. Domröse, "Direct Metal Laser Sintering," *Laser Technik Journal* **12**(1), 45-48 (2015).

[30]   E.C. Santos, M. Shiomi, K. Osakada and T. Laoui, "Rapid manufacturing of metal components by laser forming," *International Journal of Machine Tools and Manufacture* **46**(12), 1459-1468 (2006).

Chapter 2

LASERS AND BASIC OPTICS FOR 3D PRINTING AND MANUFACTURING

In this chapter, the fundamentals of light-matter interaction are discussed with illustrative examples. It begins with the description of electromagnetic radiations and the three types of phenomena that occur when light interacts with matter. This is followed by the working principle of lasers and their characteristics. The chapter concludes with collimation optics, polarisation and different but related optical components which form the basic optics for laser-assisted manufacturing and 3D printing.

2.1 Introduction

Light is an electromagnetic wave which has transverse propagation characteristics. Hence, it belongs to the family of transverse waves, like water waves, waves on a string, etc. The propagation of transverse waves can be best represented by the following schematic (see Fig. 2.1). It shows the locations of points on a string at two different times, which indicates that these points are displaced in the direction transverse to the x-direction [1].

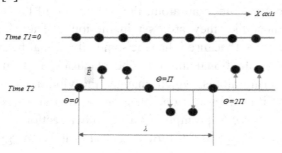

Fig. 2.1. A representation of the propagation of transverse waves.

The maximum amplitude of the oscillations is the wave amplitude—the square of which is the intensity. The length of the wave in which one full cycle of wave propagation occurs is the wavelength of the light beam. When such a light beam with a certain wavelength or a wavelength band interacts with a material surface, many phenomena can occur. The following section details these different aspects and their significance.

2.2 Interaction of Light with Materials: Different Aspects

Light can interact with matter in three different ways: absorption, transmission and reflection [2–4]. Generally, the electrons in an atom which is responsible for interactions with light, vibrate at specific natural frequencies. The interaction between the light wave frequency and the frequency of the electron's vibration determines whether the light is absorbed, reflected or transmitted.

2.2.1 *Reflection, absorption and transmission*

Reflection occurs when the frequency of the incoming light wave *does not* match that of the electrons' natural frequency. If the object is opaque (not see-through/a solid colour), the electron vibrations are not "passed down", like during absorption. Rather, the surface-level electrons vibrate briefly before emitting that wave back out (as light).

When a light wave with an identical frequency to an electron's natural frequency "impinges" upon an atom, the electrons will begin to vibrate as a result (almost like they are "set in motion"). The electrons will absorb the light wave (because it has the same vibrational frequency) and convert it into a vibrational motion. The electrons, in turn, bump up against neighbouring atoms, which changes the vibrations into thermal energy (think of it as a case of one cheering section at a baseball game starts doing the "wave"—prompting all the other section to join in the fun). This thermal energy is not turned back into light energy; thus that particular light wave never leaves the object again.

Transmission works along the same lines as reflection, except it involves transparent or semi-transparent objects. The atoms take in the wave, vibrate briefly (but at small amplitudes—not like during absorption, when they vibrate with large amplitudes), transfer the vibrations throughout the body of the material, and then re-emit the wave as light out the other end.

Fig. 2.2 depicts all the phenomena related to the interaction of light with matter.

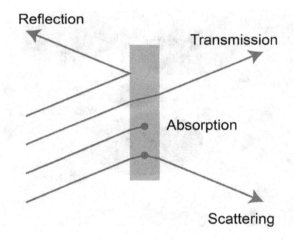

Fig. 2.2. Interaction of light with material—major phenomena.

The value of the absorpivity coefficient will vary with the same effects that affect the reflectivity.

For opaque materials,

Reflectivity = 1 − Absorptivity.

For transparent materials,

Reflectivity = 1 − (Transmissivity + Absorptivity).

In metals, the radiation is predominantly absorbed by free electrons in an "electron gas". These free electrons are free to oscillate and re-radiate without disturbing the solid atomic structure. Thus, the reflectivity of metals is very high in the waveband from the visible to very long wavelengths. As a wave-front arrives at a surface, then all the free electrons in the surface vibrate in phase, generating an electric field which is 180° out of phase with the incoming beam. The sum of this field will be a beam whose angle of reflection equals the angle of incidence.

Fig. 2.3 shows the current commercially available mirrors, reflectors and attenuators.

Fig. 2.3. Commercially available mirrors, reflectors and attenuators.

This "electron gas" within the metal structure means that the radiation is unable to penetrate metals to any significant depth beyond one to two atomic diameters. Metals are thus opaque and appear shiny. Fig 2.4 clearly illustrates the effects of reflection of light from a surface such as a mirror or a substrate surface.

The reflection coefficient for normal angles of incidence from a dielectric or metal surface in air ($n = 1$) can be calculated from the refractive index, n, and the extinction coefficient, k (or absorption coefficient as described above), for that material as follows:

$$R = [(1 - n)^2 + k^2] / [(1 + n)^2 + k^2]$$

Hence, the absorptivity for an opaque material such as a metal, A, is given by

$$A = 1 - R,$$

Where,

$$A = 4n/ [(n + 1)^2 + k^2].$$

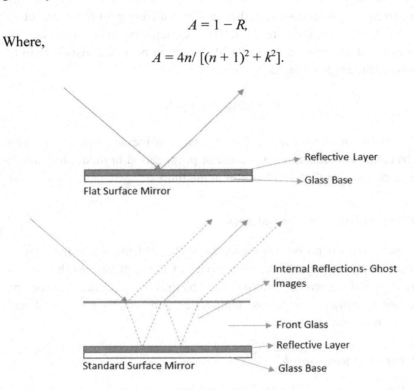

Fig. 2.4. Effects of reflection of light from a mirror substrate surface.

Further, the variation of the amplitude of the electric field, E, with respect to depth, d, is given by the Beer–Lambert law,

$$E = E_0 \exp(-2\pi kd/\lambda),$$

where, λ *is the wavelength in* vacuum [6].

The intensity is proportional to the square of the amplitude and hence the variation of intensity with depth is given by $I = I_0 \exp(-4\pi kd/\lambda)$.

2.2.2 *Refraction*

The ray undergoes refraction described by Snell's law, when the ray enters from one medium to another medium differing in their refractive indices [5]. The law indicates that, "The refracted ray lies in the plane of incidence and the sine of the angle of refraction bears a constant ratio to the sine of the angle of incidence":

$$\sin \varphi / \sin \psi = n = v_1/v_2$$

where, n is the refractive index, φ is the angle of incidence, ψ is the angle of refraction, v_1 is the apparent speed of propagation in medium 1 and v_2 is the apparent speed of propagation in medium 2.

2.3 Effect of Different Parameters

The laser's optical properties affect the surface of the workpiece where the laser beam impinges. The absorptivity of the material has the largest influence on the power requirements. Absorptivity also depends on the wavelength, surface roughness, temperature, material's phase and any use of surface coatings.

2.3.1 *Effect of wavelength*

Wavelength is defined as the characteristic spectral length associated with one cycle of vibration for a photon in the laser beam. The absorptivity of the material depends on the wavelength. Hence, certain lasers are suitable for processing only certain materials. At shorter wavelengths, the more energetic photons can be absorbed by a greater number of bound electrons and so the reflectivity falls and the absorptivity of the surface is increased.

For example, the absorption of the laser beam energy depends on both the wavelength of the laser radiation and the spectral absorptivity characteristics of the materials to be processed. Copper (Cu) and aluminium (Al) exhibit very high reflectivity to CO_2 laser radiation (10.6

µm wavelength). Nd:YAG laser radiation has high absorptivity to these metals. Hence it is more effective as the energy losses are minimal [6,7].

2.3.2 *Effect of temperature*

As the temperature of the structure rises, there will be an increase in the phonon population, inducing more phonon–electron energy exchanges. Thus, electrons are more likely to interact with the structure rather than oscillate and re-radiate. Reported research data indicate that a fall in the reflectivity and an increase in the absorptivity are expected with a rise in temperature for some metals [8–10].

2.3.3 *Effect of surface films and interference*

Reflectivity is essentially a surface phenomenon, and so surface films may have a large effect. Let us analyse a simple thin film on a substrate surface and look at what happens when light impinges on it in terms of interference phenomena [11,12] (see Fig. 2.5).

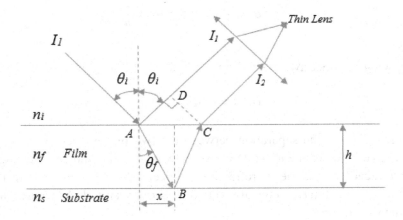

Fig. 2.5 A basic thin film interferometer.

An interferometer is an optical system that makes use of the interference phenomenon for measurement. Most of the interferometers used for technical applications either make use of a system of two surfaces, or can be reduced to one. The difference in the optical path lengths between the two interferometric arms of the interferometer is known as the optical path length difference (OPD). Fig 2.5 shows how a film on a surface acts as an interferometer. The incident light beam reaching the interface between air and film splits into two: One reflects and the other refracts into the film. The refracted beam then emanates through point C after reflecting at the bottom surface of the film (film–substrate interface). Through this process, two beams are generated and they introduce a phase difference as follows:

$$OPD = n\overline{AB} + n\overline{BC} - n_i\overline{AD}$$

$$\overline{AB} = \frac{nh}{\cos\theta_f}$$

$$\overline{ABC} = \overline{AB} + \overline{BC} = \frac{2hn}{\cos\theta_f}$$

$$\overline{AD} = 2n_i h \sin\theta_i \cos\theta_i$$

$$\sin\theta_i = \frac{\overline{AD}}{2x} \text{ and } \tan\theta_f = \frac{x}{h}$$

Applying Snell's Law: $n_i \sin\theta_i = n_f \sin\theta_f$

$$OPD = 2nh\cos\theta_f$$

where, "h" = is the separation between the two planes and "n" = is the refractive index of the medium between them. θ_i and θ_f indicate the angle of indigence and angle of refraction, respectively. At points where the conditions of interference are satisfied, bright and dark fringes are formed respectively.

A fringe passes through all those points where the path difference is a constant. The path difference introduced between the two waves should be less than the coherence length (the length of the wave train emitted by the light source) to produce the interferences. When the path difference is large, waves do not superpose and no interference is produced [13].

For constructive interference,

$$2\,n\,h\cos\theta_f = \frac{2\,m\,\lambda}{2}$$

For destructive interference,

$$2\,n\,h\cos\theta_f = \frac{(2\,m+1)\,\lambda}{2}$$

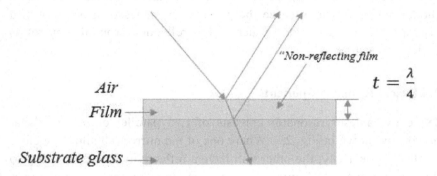

Fig. 2.6. Minimum thickness of coated film on a substrate glass for antireflection.

Notice that there are no phase shifts due to reflections in this case. The minimum film thickness for the destructive interference of the reflected light to occur is as follows:

$$t_{min} = \frac{\lambda_{film}}{4}$$

Fig. 2.6 shows that for interference coupling the film must have a thickness of around $[(2n+1)/4]\,\lambda$ to have any effect, where n is any integer. The generic absorption variation for CO_2 laser radiation by a surface oxide film due to such a reason is shown. A form of these surface films may be plasma [14] provided that the plasma is in thermal contact with the surface.

2.3.4 *Effect of materials and surface roughness*

Roughness has a large effect on absorption owing to the multiple reflections due to randomisation. There may also be some "stimulated absorption" due to beam interference with sideways-reflected beams [14]. If the roughness is less than the beam wavelength, the radiation will not suffer these events and hence will perceive the surface as flat. The reflected phase front from a rough surface, formed from the Huygens wavelets, will no longer be the same as the incident beam and will spread in all directions as a *diffuse reflection*. It is interesting to note that it should not be possible to see the point of incidence of a beam from a laser pointer on a mirror surface if the reflection from the mirror is perfectly specular [15].

2.4 Laser: Basic Components

Typically, a laser resonator consists of two parallel or nearly planar mirrors as shown in Fig. 2.7. Where one of the mirrors should have close to 100% reflectivity, the other will have a reflectivity less than 100% so that some of the light from the resonator can transmit through this mirror, which will act as the window, and thus form the laser beam [16].

Light travels back and forth between the mirrors multiple times. They interfere constructively if the following equation is satisfied.

$$m\lambda_0 = 2L \cos\theta$$
$$m\lambda_0 = 2L \; for \quad \theta \cong 0^0$$
$$m(\lambda_0 / 2) = L$$

Since light that travels perpendicular to the resonator mirrors will remain within the resonator, they will be significantly amplified. This will make a standing wave pattern between the to and fro travelling waves, and thus form standing wave patterns (modes) which are governed by the following equations,

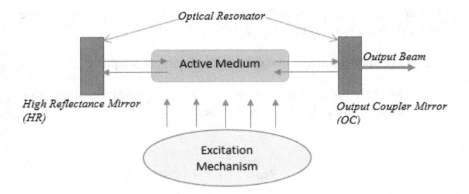

Fig. 2.7. Basic components of a laser resonator.

If L is the cavity length, then L will be equal to an integral number of half-wavelengths for the laser to operate at the certain specific wavelength. In terms of frequency, since $v = c/v_0$, the laser mode frequency is given by, $v_m = mc/2L_c$. These modes will have a separation between them with a frequency difference of $\Delta v = c/2L$, as shown in Fig. 2.8, where c is the velocity of light in vacuum. So the spectrum of light of a laser looks like a series of narrow peaks spaced by the mode spacing.

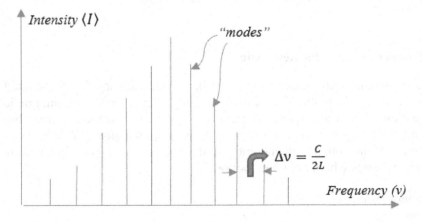

Fig. 2.8. Longitudinal modes.

Most laser systems use curved mirrors instead of true planar mirrors due to the following reasons (see Fig. 2.9):

(a) Curved mirrors make the laser alignment simple.

(b) Due to its inherent nature of accommodating diffraction effects, cavity losses are minimised.

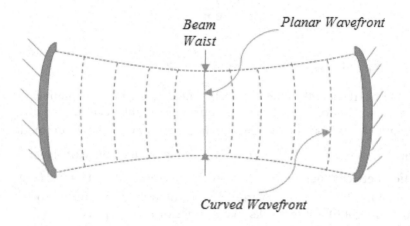

Fig. 2.9. Curved resonators in a laser.

2.5 Laser Beam: Characteristics

Laser, which is the short form for Light Amplification by Stimulated Emission of Radiation, is a high-energy beam of electromagnetic radiation. The light (photon) travels as a wave through space, but behaves as a particle of energy when it encounters matter [17]. The laser differs from other incoherent light sources due to its unique characteristics which are given below,

• monochromatic
• coherent
• directional or collimated
• bright

2.5.1 *High monochromaticity*

The laser beam can be considered to be monochromatic because the oscillating laser consists of very closely spaced, discrete and narrow spectral lines (laser modes or cavity modes) compared with a conventional light source whose emission covers a frequency bandwidth in the order of gigahertz. For better monochromaticity, the single mode can be achieved by forcing the laser to oscillate in a single transverse mode (usually, the fundamental TEM00 Gaussian mode and longitudinal mode).

The laser is light that has nearly a single colour (due to stimulated emission) in contrast with non-laser light sources (e.g. thermal sources) which produces light mainly through spontaneous emission and gives out polychromatic (near-white) light (See Fig. 2.10).

The reason for the laser to be so monochromatic is due to the fact that the gain occurs at a well-defined frequency determined by the transition frequency of the atoms in the gain medium [18].

Also, the laser light, being a stimulated emission originating from a few similar photons, can be much narrower than the spontaneous emission from a single transition.

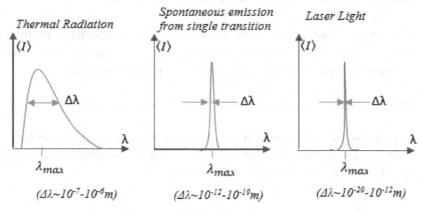

Fig. 2.10. Spectral bandwidth comparison.

2.5.2 *Extreme directionality*

Since almost the entire laser light is a stimulated emission that originates from a few photons and travels along the laser axis, the laser beam is essentially perfectly collimated (all rays are parallel). The laser beam is directional or collimated due to its small divergence angle ranging from 0.2 to 10 milliradians, except for semiconductor laser. The directionality of a laser beam enables it to be focused on a very small spot over a long distance [18].

2.5.3 *High brightness*

Spectral brightness plays a very important role when handling laser sources for many applications. Just to give the significance of this term, let us estimate the spectral brightness of the low power He-Ne laser. It can be found that its spectral brightness is 10,000,000,000 times higher than that of the sun. For a 1 mW He-Ne laser, it is found to be 5×10^{23} W/(m^3-sterad) [19].

2.5.4 *Extreme coherence*

Coherence is the measure of the correlation between the phases measured at different points on a wave. Though it is a property of a propagating wave, coherence is directly related to the wave source's characteristics.

The laser beam is coherent because of the fixed-phase relationship between two waves at the wavefront over time (spatial coherence) or between two points of the same wave (temporal coherence).

The two basic types of coherence [20,21] are described as follows:

(a) Temporal coherence: It is a measure of the correlation between the light waves' phases at different points along the propagation direction. It implies how monochromatic a light source is!

(b) Spatial coherence: It is a measure of the correlation of the light waves' phases at different points transverse to the propagation direction. It implies how uniform the phase of the wavefront is.

For example: see Fig. 2.11 (a). It is a very incoherent incandescent bulb. We can always make an incoherent source coherent as shown in Fig. 2.11 (b) by throwing away some light (loss of light!). Spatial filtering is done to increase the spatial coherence followed by spectral filtering to increase the temporal coherence. But, the laser is naturally very coherent.

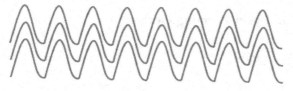

(a) Coherent Light

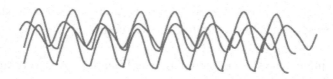

(b) Incoherent Light

Fig. 2.11. Coherent and incoherent light beams.

2.5.5 *Directionality*

The laser beam is highly directional, which implies laser light has a very small divergence. This is a direct consequence of the fact that the laser beam comes from the resonant cavity, and only waves propagating along the optical axis can be sustained in the cavity. The directionality is described by the light beam divergence angle. Please see Fig. 2.11 to observe the relationship between divergence and optical systems.

For perfect spatial coherent light, a beam of aperture diameter D will have unavoidable divergence because of diffraction [22]. From diffraction theory, the divergence angle q_d is:

$$q_d = b_l/D$$

where, b_l and D are the wavelengths and the diameter of the beam respectively. b is a coefficient whose value is around unity and depends on the type of light amplitude distribution and, the definition of the beam's diameter. q_d is called the diffraction limited divergence.

2.6 Laser Beam: Beam Forming Optics

2.6.1 *Collimation optics*

A **collimator** is a device that narrows a beam of particles or waves. It is a device that filters a stream of rays so that only those travelling parallel to a specified direction are allowed through. To narrow can mean either to cause the directions of motion to become more aligned in a specific direction (i.e., make **collimated** light or parallel rays), or to cause the spatial cross section of the beam to become smaller.

There are many ways of implementing collimation. A diverging beam can be collimated if one keeps a lens at its focal plane from the origin of the source (see Fig. 2.12).

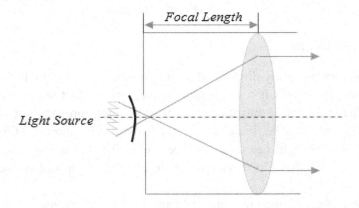

Fig. 2.12. Collimation Principle.

Collimated light is light whose rays are parallel, and therefore will spread minimally as it propagates. A perfectly collimated beam, with no divergence, would not disperse with distance. Such a beam cannot be created, due to diffraction. Light can be approximately collimated by a number of processes, for instance by means of a collimator. Perfectly collimated light is sometimes said to be *focused at infinity*. Thus as the distance from a point source increases, the spherical wavefronts become flatter and closer to plane waves, which are perfectly collimated.

2.6.2 *Polarisation*

In an electromagnetic wave, both the electric field and magnetic field are oscillating but in different directions; by convention, the "polarisation" of light refers to the polarisation of the electric field. The light which can be approximated as a plane wave in free space or in an isotropic medium propagates as a transverse wave—both the electric and magnetic fields are perpendicular to the wave's direction of travel. The oscillation of these fields may be in a single direction (linear polarisation), or the field may rotate at the optical frequency (circular or elliptical polarisation). In that case, the direction of the fields' rotation, and thus the specified polarisation, may be either clockwise or counter-clockwise; this is referred to as the wave's chirality or *handedness* [23] (see Fig. 2.13.).

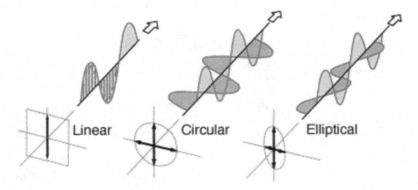

Fig. 2.13. The different states of polarisation of a beam.

The most common optical materials (such as glass) are isotropic and simply preserve the polarisation of a wave but do not differentiate between polarisation states. However, there are important classes of materials classified as birefringent or optically active in which this is not the case and a wave's polarisation will generally be modified or will affect propagation through it. A polariser is an optical filter that transmits only one polarisation.

2.6.3 *Polarisers and phase plates*

A polariser acts as an optical filter that only allows light of a specific polarisation to pass through (blocking the rest) (see Fig. 2.14.). It can thus convert a light beam of mixed/undefined polarisation into a polarised light beam, which would have a well-defined polarisation. The most common types of polarisers are linear and circular polarisers.

The propagation of photons in space is characterised by its oscillatory motion in its electrical field vector, to produce an electromagnetic wave. This wave is representative of the direction and magnitude of the photon's electric field vector as a function of time. Due to the high coherence nature of a laser beam, all the photons in the beam are aligned in the same direction and thus the laser beam can be defined as linearly polarised. This property of the state of polarisation of the beam has a

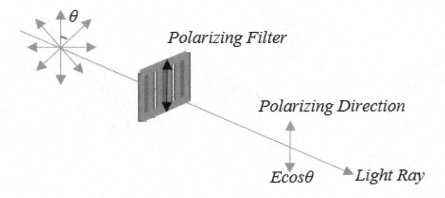

Fig. 2.14. Linear polariser.

certain influence on the cutting and grooving operations using lasers [24,25].

The two different categories of linear polarisers are, absorptive and beam splitting polarisers. In an absorptive polariser, all the unwanted polarisation states are absorbed, while for a beam splitting polariser, the unpolarised beam is divided into two with opposite polarisation states.

Malus' law says that when a perfect polariser is placed in a polarised beam of light, the intensity, I, of the light that passes through is given by [26]

$$I = I_0 \cos^2 \theta_i$$

where I_0 is the initial intensity, and θ_i is the angle between the light's initial polarisation direction and the axis of the polariser.

2.6.4 *Beam splitters*

A beam splitter is an optical device that splits a beam of light into two (see Fig. 2.15.). They are widely used in most of the interferometers and form a crucial segment. For machining and material processing applications, similar efficiencies can be achieved by using a laser beam that is split into two, compared to using two laser beams of the same coherence properties. The most common form of a beam splitter is a cube that is made of two triangular glass prisms glued together at their base. Polyester, epoxy or urethane-based adhesives are a few of the common glues used for this process. The thickness of the resin layer is adjusted such that for a certain wavelength of light used, half of the incident light through one "port' (i.e. the face of the cube) is reflected while the rest is transmitted. This happens due to frustrated total internal reflections of the light in the thin epoxy layer. Polarising beam splitters, such as the Wollaston prism, split the light beam into two beams of different polarisations. This is achieved by the use of birefringent materials in its construction.

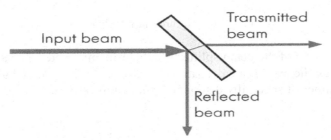

Fig. 2.15. A 50:50 beam splitter configuration.

Other possible designs of a beam splitter use a half-silvered mirror, a sheet of glass, or plastics with a transparently thin coating of metal (now usually aluminium deposited from aluminium vapour). The thickness of the deposit is controlled such that a part of the incident light (at a 45-degree angle) which is not absorbed by the coating, is transmitted. The remainder of the light is thus reflected. In photography, such thin half-silvered mirrors known as "pellicle mirror" are used. The so-called "Swiss cheese" beam splitter mirrors are used to reduce the loss of light due to absorption by a reflective coating. The original design of these mirrors used sheets of polished metal perforated with holes—the number and distribution of which depended on the requirement of the desired ratio of reflection to transmission. Further advancements included the use of discontinuous coatings resulting from the metal being sputtered onto the glass or the removal of small areas of continuous coating using chemical or mechanical means. The produced surface can be literally called a "half-silvered" surface.

Instead of a metallic coating, a dichroic optical coating may be used (see Fig. 2.16.). Depending on its characteristics, the ratio of reflection to transmission will vary as a function of the wavelength of the incident light. Dichroic mirrors are used in some ellipsoidal reflector spotlights to split off unwanted infrared (heat) radiation and as output couplers in laser construction [27,28].

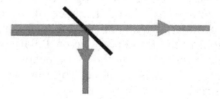

Fig. 2.16. A dichroic beam splitter.

A third version of the beam splitter is a dichroic mirrored prism assembly which uses dichroic optical coatings to divide an incoming light beam into a number of spectrally distinct output beams (see Fig. 2.17).

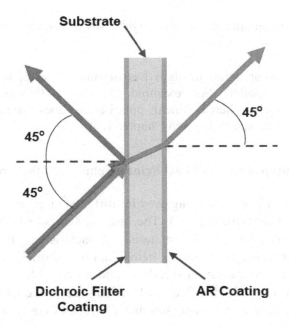

Fig. 2.17. Working principle of a dichroic beam splitter.

2.6.5 *Wave plates*

A wave plate or a retarder is an optical device that alters the states of polarisation of the light beam passing through it. The two most common types of wave plates are the half wave plate and the quarter wave plate. A half wave plate shifts the polarisation states of linearly polarised light and a quarter wave plate converts linearly polarised light into circularly polarised light and vice versa [29,30].

Wave plates are constructed out of birefringent materials (such as quartz or mica). Birefringent materials have a different index of refraction in different orientations with respect to the light passing through it. The behaviour of a wave plate, be it half wave or a quarter wave, depends on the crystal's thickness, the wavelength of the light used, and the variation of index of refraction. By the appropriate choice of the mentioned factors, it is possible to introduce a controlled phase shift between the two polarisation components of the light wave, thereby altering the polarisation.

The state of polarisation of light has detrimental effects when light interacts with matter. For example, circular polarisation has no directional effects whereas linear polarisation does affect. Detailed explanations of these are given in Chapter 4.

2.7 Representative Lasers in 3D Printing and Manufacturing

Various kinds of lasers are being used for different target materials in 3D printing and manufacturing [31]. The laser system should be carefully selected by considering the material characteristics, for example, absorption wavelength, melting temperature, thermal conductivity, thermal capacitance, ablation threshold and so on [32,33]. Here, four representative lasers in this field will be introduced: the CO_2 gas laser, Nd:YAG crystal laser, Excimer laser and Yb-doped fibre laser [34–36].

2.7.1 CO_2 laser

CO_2 laser is one of the most widely used in 3D printing [37–39]. It is a molecular gas laser that produces light through ionisation by electrical discharge. There are three basic components in a CO_2 laser, namely, the gain medium, pump source and cavity mirrors. As shown in Fig. 2.20, the CO_2 laser consists of a gas tube (as the gain medium), an electrical pump and cavity mirrors (high-reflectivity mirror and output coupler). If the average power level increases, additional cooling elements should be installed; passive air-cooling, active forced air-cooling, thermo-electric cooling and forced water-cooling are examples. In the gas tube working as the gain medium, the gas mixture of CO_2, He and N_2 is filled in and then electrically pumped. Its output wavelength is at 10.6 μm infrared (IR) range, which is well suited for polymers due to their strong IR absorption. Table 2.1 shows the specification example of a commercially available CO_2 laser.

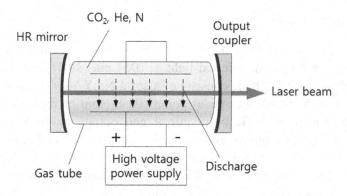

Fig. 2.20. Typical configuration of a (sealed tube) CO_2 laser.

Table 2.1. Example specification of a CO_2 laser.

CO_2 laser	
Average Output Power	400 W
Wavelength	10.6 μm
Rise Time / Fall Time	< 100μs / < 100μs
Power Stability from Cold Start	± 7%
Power Stability after 3 Minutes (typical)	± 5%
Duty Cycle Range	1% - 100%
Operating Frequency	0–100 kHz
Beam Waist Diameter (at $1/e^2$)	6.0 mm ± 0.6 mm
Mode Quality	$M^2 \leq 1.2$
Polarisation	Linear (45 degrees)
Cooling	Water
Input Voltage/Current (maximum)	48 V_{DC} / 125 A

2.7.2 *Nd:YAG laser*

A Nd:YAG ($Nd^{3+}:Y^3Al^5O^{12}$) laser is a solid-state laser (different from a gas laser) emitting near-infrared light at 1064 nm [40,41]. As shown in Fig. 2.21, a Nd:YAG laser consists of a solid Nd:YAG synthetic crystal (as the gain medium), an optical pump (flash-lamp or diode laser) and cavity mirrors (high-reflectivity mirror and output coupler). Nd:YAG lasers can support continuous-wave (CW), long-pulsed and short-pulsed modes. For pulsed modes, highly-doped Nd:YAG crystals need to be used for higher gain factors within a relatively short resonating time. For relatively long pulse generation (longer than 0.2 ms), the pump modulation (e.g. modulation of flash lamps) is good enough. For shorter pulses (e.g. pulse durations of tens of ns), an intra-cavity optical switch—the fourth component in a pulsed laser—is required for gain or loss modulation. These optical switches work to maximise the population inversion state such that short laser pulses can be generated; this is called as the Q-switching technique. With the aid of Q-switching, the pulse duration of the output beam can be shortened from hundreds of microseconds to tens of nanoseconds. Thanks to this short pulse duration, nonlinear wavelength conversion can be realised with high conversion efficiency. Thus, the second harmonic generation and third harmonic generation convert the fundamental wavelength (1064 nm) to 532 nm and 355 nm, respectively. More recently, Nd:YVO$_4$ lasers have been replacing the position of Nd:YAG lasers, because of their wider absorption wavelength bandwidth, lower lasing threshold and higher efficiency. For an example Nd:YVO$_4$ laser specification, refer to Table 2.2.

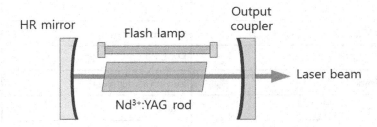

Fig. 2.21. Typical configuration of a Nd:YAG laser.

Table 2.2. Example specification of a Nd:YAG laser.

Nd:YAG laser model	
Average Output Power	100 W
Wavelength	1064 nm
Power at 10 kHz (W)	75
Power at 6 kHz (W)	60
Pulse-to-Pulse Stability (% RMS)	<2
Pulsewidth	< 160 ns
Beam Pointing Stability	< 20
Beam Diameter (mm)	6 mm
Beam Divergence	9.5 mrad
Beam Quality (M^2)	< 20
Polarisation	random

2.7.3   *Excimer laser*

Excimer lasers emit repetitive nanosecond pulses in a short ultraviolet wavelength range [42–45]. 'Excimer' is the abbreviation of 'excited dimer'. It indicates that the gas mixture contains a noble gas (e.g. Argon, Xenon or Krypton), a halogen (Fluorine or Chlorine), and a buffer gas (Neon or Helium). Depending on the gas composition, the output wavelength can be varied from 157 nm to 351 nm. The three basic components of the laser are thus the excimer (working as the gain material), electrical pumping, and two UV cavity mirrors (used for the resonator). The repetition rate is about a few kHz and the output average power is up to a few kilowatts. Although excimer lasers provide a relatively poor beam quality at an extremely high price, they still hold unique positions in the market because there are no comparative high power laser sources in the ultraviolet wavelength region. To improve the patterning resolution in semiconductor lithography, the wavelength of the patterning laser should be short because of the optical diffraction limit, which is why excimer lasers are widely used in semiconductor industries. Table 2.3 shows an example specification of an Excimer laser.

Table 2.3. Example specification of an Excimer laser.

Excimer laser	
Average Output Power	1200 W
Wavelength	308 nm
Pulse duration (FWHM)	24 ± 4 ns
Max Pulse Energy	2000 mJ
Max Repetition Rate	600 Hz
Beam Dimensions (mm^2)	35 ± 4 x 14.5 ± 3
Beam Divergence (mrad^2)	4.5×1.3

2.7.4 *Yb-doped fibre laser*

A Yb-doped fibre laser is one of the rare-earth-doped fibre lasers that can be used as the gain material for high-power lasers [46–48]. Among the rare-earth-doped materials, Ytterbium (Yb) has small quantum defects and so can potentially attain very high power efficiencies without thermal side effects. Because Yb-doped fibre laser use Yb-doped optical fibre as the gain medium and is pumped by fibre-coupled diode lasers, it can operate with much higher laser gain and lower resonator losses. Fibre lasers nowadays support very high-power outputs with average powers exceeding hundreds of watts and even several kilowatts from a single fibre (several single-mode fibres can be used in parallel). This potential arises from a high surface-to-volume ratio (enabling efficient thermal cooling) and the guiding effect (with minimal energy loss), which avoids thermo-optic problems. Laser cavity structures can differ depending on the output power level. For stable low-power operation, the laser oscillator based on the ring-cavity is better (see Fig. 2.22) whereas the linear cavity design is advantageous for higher power operations. Refer to Table 2.4 for an example specification.

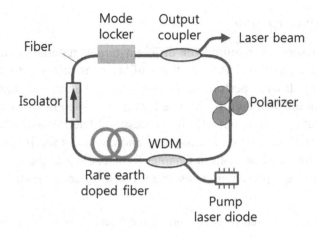

Fig. 2.22. Typical configuration of a Yb-doped fibre laser.

Table 2.4. Example specification of a Yb-doped fibre laser.

Yb-doped fibre laser	
Average Output Power	100 W
Wavelength	1030–1070 nm
Central Wavelength Accuracy	± 1 nm
Mode of Operation	CW/Modulated
Power Tunability	10–100 %
Power Stability	± 2 %
Optical Noise	< 2 % RMS
Polarisation	Linear, 50:1
Beam Quality (M^2)	<1.1

2.7.5 *Laser comparison*

Different lasers are required for different target materials and different manufacturing methods. As new types of lasers are coming to the market and existing lasers' performances are getting better, the laser system adopted should be carefully selected with an in-depth understanding of the 3D printing and manufacturing processes. Material-dependent optical properties will be covered in Chapter 3 and the lasers for 3D printing and manufacturing will be dealt in Chapter 5. In Table 2.5, the specifications of four existing representative lasers are listed for comparison.

Table 2.5. Specification comparison of different lasers in 3D printing.

Laser	CO_2 laser	Nd:YAG laser	Excimer laser (Xenon fluoride)	Yb-fibre laser
Application	SLA, SLM, SLS, LENS	SLM, SLS, LENS	SLA	SLM, SLS, LENS
Operation wavelength	0.6 µm	1.064µm	0.35 µm	1.07 µm
Pump source	Electrical discharge	Flash lamp, laser diode	Electrical discharge	Laser diode
Operation mode	CW & Pulse	CW & Pulse	Pulse	CW & Pulse
Pulse duration	Hundreds ns-tens µs	Few ns– tens ms	Tens ns	Tens ns– tens ms
Beam quality factor (mm·mrad)	3–5	0.4–20	160 × 20 (Vertical × Horizontal)	0.3–4
Fibre delivery	Not possible	Possible	Possible	Possible
Maintenance periods (hrs)	2000	200 (lamp life)	10^{8-9} pulses (thyratron life)	Maintenance free

2.8 Summary

The fundamentals of light-matter interaction have been detailed with representative schematic diagrams. A thin film interferometer has been illustrated and the required optimum film thickness for fabricating anti-reflection coatings explained. Laser characteristics and their effect on machining have also been analysed. Finally, all the important optical components and their working details have been discussed and analysed. These concepts and details are significant while designing optics for laser assisted manufacturing and 3D printing.

Problems

1. With the help of a schematic diagram, explain briefly the different phenomena that occur when light interacts with matter.

2. Derive the formula to represent the intensity pattern of the interfered beams in a thin film interferometer.

3. Derive the relationship between the minimum thicknesses of the film to be coated on a glass surface for it to act as an antireflection film.

4. How can one obtain circular polarisation of light beam?

5. What is Malus' law? Draw the optical configuration to demonstrate Malus' law and mark individual optical components.

6. Explain coherence property associated with laser beams. Which are the two different types of coherence?

7. What is a dichroic beam splitter? Explain its working principle and how it differs from normal cube beam splitter.

8. What are the different laser beam characteristics?

9. Why curved mirrors are preferred in laser resonators? Explain with a brief schematic diagram.

10. How surface roughness of a material affects light absorption? What is meant by diffusivity?

11. List four representative lasers for 3D printing and manufacturing.

a.   What are the maximum powers of the four lasers described?

b.   What are the central wavelengths of these lasers?

c.   What are the repetition rates of them?

d.   Discuss the maintenance issues.

References

[1]   Snyder, A.W. and J. Love, *Optical waveguide theory*. 2012: Springer Science & Business Media.

[2]   Auld, B.A., *Acoustic fields and waves in solids*. 1973: Рипол Классик.

[3]   Taylor, J.R., *Optical solitons: theory and experiment*. Vol. 10. 1992: Cambridge University Press.

[4]   Yasumoto, K., *Electromagnetic theory and applications for photonic crystals*. 2005: CRC press.

[5]   Shirley, J.W., *An early experimental determination of Snell's law*. American Journal of Physics, 1951. **19**(9): p. 507–508.

[6]   Lu, J., *et al.*, *Optical properties and highly efficient laser oscillation of Nd:YAG ceramics*. Applied Physics B, 2000. **71**(4): p. 469–473.

[7]   Österberg, U. and W. Margulis, *Dye laser pumped by Nd:YAG laser pulses frequency doubled in a glass optical fiber*. Optics Letters, 1986. **11**(8): p. 516–518.

[8]   Winsemius, P., et al., *Temperature dependence of the optical properties of Au, Ag and Cu*. Journal of Physics F: Metal Physics, 1976. **6**(8): p. 1583.

[9]   Bennett, H., M. Silver, and E. Ashley, *Infrared reflectance of aluminum evaporated in ultra-high vacuum*. JOSA, 1963. **53**(9): p. 1089–1095.

[10]   Smith, D.R. and F. Fickett, *Low-temperature properties of silver*. Journal of Research-National Institute of Standards And Technology, 1995. **100**: p. 119–119.

[11]   Heavens, O.S., *Optical properties of thin solid films*. 1991: Courier Corporation.

[12]   Hass, G., M.H. Francombe, and R.W. Hoffman, *Physics of Thin Films: Advances in Research and Development*. 2013: Elsevier.

[13]   Rancourt, J.D., *Optical thin films: user handbook*. 1996: SPIE Press.

[14]   Sandhyarani, M., *et al., Surface morphology, corrosion resistance and in vitro bioactivity of P containing ZrO 2 films formed on Zr by plasma electrolytic oxidation*. Journal of Alloys and Compounds, 2013. **553**: p. 324–332.

[15]   Li, J.-M., *et al., Self-absorption reduction in laser-induced breakdown spectroscopy using laser-stimulated absorption*. Optics letters, 2015. **40**(22): p. 5224–5226.

[16]   Haken, H., *Laser theory*. 2012: Springer Science & Business Media.

[17]   Newton, R.G., *Scattering theory of waves and particles*. 2013: Springer Science & Business Media.

[18]   Davis, C.C., *Lasers and electro-optics: fundamentals and engineering*. 2014: Cambridge University Press.

[19]   Jamil, Y., *et al., He–Ne laser-induced changes in germination, thermodynamic parameters, internal energy, enzyme activities and physiological attributes of wheat during germination and early growth*. Laser Physics Letters, 2013. **10**(4): p. 045606.

[20]   Dainty, J.C., *Laser speckle and related phenomena*. Vol. 9. 2013: Springer Science & Business Media.

[21]   Ackermann, S., *et al., Generation of coherent 19-and 38-nm radiation at a free-electron laser directly seeded at 38 nm*. Physical review letters, 2013. **111**(11): p. 114801.

[22]   Löffler, W., A. Aiello, and J. Woerdman, *Spatial coherence and optical beam shifts*. Physical review letters, 2012. **109**(21): p. 213901.

[23]   Clarke, D. and J.F. Grainger, *Polarized Light and Optical Measurement: International Series of Monographs in Natural Philosophy*. Vol. 35. 2013: Elsevier.

[24]   Menzel, R., *Photonics: linear and nonlinear interactions of laser light and matter*. 2013: Springer Science & Business Media.

[25]   Chryssolouris, E., *Laser machining: theory and practice*. 2013: Springer Science & Business Media.

[26]   Glasenapp, P., *et al., Resources of polarimetric sensitivity in spin noise spectroscopy*. Physical Review B, 2013. **88**(16): p. 165314.

[27]   Rowlands, C.J., *et al. Near-Infrared Temporal Focusing Microscopy with Quantum Dot Fluorophores.* in *Optical Tomography and Spectroscopy.* 2016. Optical Society of America.

[28]   Zhao, Z., *et al.*, *High-power fiber lasers for photocathode electron injectors.* Physical Review Special Topics-Accelerators and Beams, 2014. **17**(5): p. 053501.

[29]   Xie, Y.-Y., *et al.*, *Simple method for generation of vector beams using a small-angle birefringent beam splitter.* Optics letters, 2015. **40**(21): p. 5109–5112.

[30]   Yu, N., *et al.*, *A broadband, background-free quarter-wave plate based on plasmonic metasurfaces.* Nano letters, 2012. **12**(12): p. 6328–6333.

[31]   Chua, C. K., & Leong, K. F. (2015). 3D printing and additive manufacturing: principles and applications: principles and applications. World Scientific, Singapore.

[32]   Kruth, J. P., Leu, M. C., & Nakagawa, T. (1998). *Progress in additive manufacturing and rapid prototyping.* CIRP Annals-Manufacturing Technology, **47**(2), 525–540.

[33]   Vaezi, M., Seitz, H., & Yang, S. (2013). *A review on 3D micro-additive manufacturing technologies.* The International Journal of Advanced Manufacturing Technology, **67**(5–8), 1721–1754.

[34]   Yariv, A. (1976). Introduction to optical electronics.

[35]   Saleh, B. E., Teich, M. C., & Saleh, B. E. (1991). *Fundamentals of photonics* (Vol. 22). New York: Wiley.

[36]   Paschotta, R. (2008). Encyclopedia of laser physics and technology. Berlin: Wiley-vch.

[37]   Beaulieu, A. J. (1970). *Transversely excited atmospheric pressure CO_2 lasers.* Applied Physics Letters, **16**(12), 504–505.

[38]   Wood, O. R., & Schwarz, S. E. (1967). *Passive Q-switching of a CO_2 laser.* Applied Physics Letters, 11(3), 88–89.

[39]   Nayak, N. C., Lam, Y. C., Yue, C. Y., & Sinha, A. T. (2008). *CO_2-laser micromachining of PMMA: the effect of polymer molecular weight.* Journal of Micromechanics and Microengineering, **18**(9), 095020.

[40]   Konno, S., Fujikawa, S., & Yasui, K. (1997). 80 W cw TEM00 1064 nm beam generation by use of a laser-diode-side-pumped Nd:YAG rod laser. Applied Physics Letters, **70**, 2650–2651.

[41]   Lu, J., Prabhu, M., Song, J., Li, C., Xu, J., Ueda, K., ... & Yanagitani, T. (2000). *Optical properties and highly efficient laser oscillation of Nd: YAG ceramics.* Applied Physics B, **71**(4), 469–473.

[42]   Kawamura, Y., Toyoda, K., & Namba, S. (1982). *Deep uv submicron lithography by using a pulsed high-power excimer laser.* Journal of Applied Physics, **53**(9), 6489–6490.

[43]   Dyer, P. E. (2003). *Excimer laser polymer ablation: twenty years on.* Applied Physics A, **77**(2), 167–173.

[44]   Trokel, S. L., Srinivasan, R., & Braren, B. (1983). *Excimer laser surgery of the cornea.* American Journal of Ophthalmology, **96**(6), 710–715.

[45]   Ihlemann, J., Wolff, B., & Simon, P. (1992). *Nanosecond and femtosecond excimer laser ablation of fused silica.* Applied Physics A, **54**(4), 363–368.

[46]   Jeong, Y. E., Sahu, J. K., Payne, D. N., & Nilsson, J. (2004). *Ytterbium-doped large-core fiber laser with 1.36 kW continuous-wave output power.* Optics Express, **12**(25), 6088–6092.

[47]   Limpert, J., Schreiber, T., Clausnitzer, T., Zöllner, K., Fuchs, H., Kley, E., ... & Tünnermann, A. (2002). *High-power femtosecond Yb-doped fiber amplifier.* Optics Express, **10**(14), 628–638.

[48]   Cheo, P. K., Liu, A., & King, G. G. (2001). *A high-brightness laser beam from a phase-locked multicore Yb-doped fiber laser array.* IEEE Photonics Technology Letters, **13**(5), 439–441.

Chapter 3

MATERIALS FOR LASER-BASED 3D PRINTING AND MANUFACTURING

For realising higher resolution and higher productivity in laser-based 3D printing and manufacturing over various materials, photon interaction with target materials should be carefully considered. Materials for 3D printing and manufacturing are categorised into different ways depending on their applications, but generally can be categorised as either polymers, metals or ceramics. In the field of 3D printing, materials are better to be categorised by their original form as solid-, liquid- or powder-based ones. In this chapter, various materials in laser-based 3D printing and manufacturing will be discussed for the best quality material rocessing.

3.1. Introduction

Lasers are one of the most important sources of energy in material processing due to several reasons [1–3]. Firstly, they provide high energy transfer efficiency. The laser beam can propagate over millions of kilometres with minimal energy loss thanks to its high spatial coherence; then, the laser beam can be focused into a focal spot using refractive or reflective optics with high efficiency. However, mechanical, thermal and electrical energy transfer processes will suffer from severe energy loss during energy transfer or concentration. Secondly, a laser beam can be focused onto a small spot. The minimal spot size is fundamentally limited by the diffraction limit, which is roughly half of the laser wavelength in use (~500 nm in the case of a 1060-nm Yb-fibre laser). Therefore, in principle, sub-micrometre-level, high-resolution material processing can be realised. Thirdly, the lasers can support high-speed material processing. Due to lasers' high energy density, the material temperature can be instantaneously increased over thousands of degrees

Celsius within an exposure time of a millisecond; so a laser beam can efficiently meltdown polymers, metals and some ceramics. If short-pulse lasers are used, the material temperature can be much more drastically increased. Fourthly, materials can be selectively processed due to their wavelength dependent absorption. By carefully selecting the laser wavelength and considering the material absorption, the target material can be efficiently processed without affecting side effects to other materials located nearby, underneath or over the target material. These critical advantages are the reasons why lasers are widely used today in 3D printing and manufacturing [4,5].

While there are many ways in which one can classify the numerous 3D printing and manufacturing systems in the market, materials can be generally categorised in three groups: polymers, metals and ceramics. One of the better ways, especially in 3D printing, is to classify the materials by the initial form of its base material. In this manner, all manufacturing system can be easily categorised as (1) solid-, (2) liquid- or (3) powder-based ones as shown in Fig. 3.1 [6]. Compared with

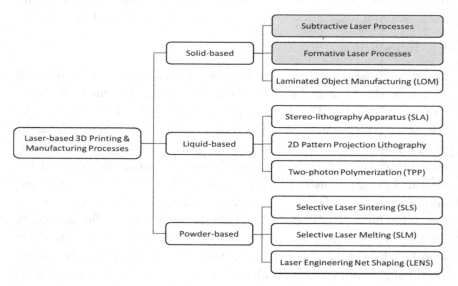

Fig. 3.1. Laser-based 3D printing and manufacturing processes are categorised by their raw material phase: solid-, liquid- and powder-based ones.

subtractive and formative manufacturing, additive manufacturing is more sensitive to the forms of base materials.

3.2. Materials for 3D Printing and Manufacturing

Material selection is one of the most critical parts of the manufacturing process because it affects the general properties of product, such as its mechanical, thermal, electrical and optical properties. Once a suitable material is chosen, only then can the appropriate manufacturing process can be selected by considering product cost, labour, time, equipment other technological details. Because material cost could make up more than 50% of the total manufacturing cost, the material selection process becomes much more critical [7]. This chapter explains basic material selection methods, fundamental base material properties, such as mechanical, thermal, electrical properties and their optical properties, to help readers better understand the laser–material interaction.

There are various concerns in material selection for laser-based 3D printing [8]. Material properties can be regarded as the fundamental link between the base material and the product's performance. Therefore, various material properties must be carefully considered from the beginning [9]. Depending on material properties, the manufacturing process will be determined. Here is a list of questions which can be raised in the material selection process.

What are the most critical objectives of the product?
- *Low price*
- *Less weight*
- *High strength*
- *High conductivity*
- *High viscosity*
- *High reflectance*
- *High damage threshold*

- *Low melting temperature*
- *Environmental stability*
- *Combination of these*

Which are the material candidates for satisfying these   objectives?
- *Metal*
- *Polymer*
- *Ceramic*
- *Combination of these*

How can we manufacture the products?
- *Subtractive process*
- *Formative process*
- *Additive process*

What fundamentally limits the production process?
- *Time*
- *Production cost*
- *System complexity*
- *Post processing*

Do we have alternative materials for better product or processing?

Table 3.1 shows basic material properties which need to be considered in laser-based 3D printing and material processing. Optical properties are important for selecting an appropriate laser system, beam delivery optics and manufacturing process. Meanwhile, mechanical, thermal, chemical properties and environmental sensitivities determine the performance of the final product. Since there are a large number of materials, one should sort out improper candidates and select appropriate ones by considering users' critical design concerns.

Most materials in 3D printing and manufacturing can be categorised into three basic ones: polymers, metals and ceramics, as shown in Fig. 3.2.

Table 3.1. Material properties that need to be considered in laser-based manufacturing.

	Material properties
Optical properties	Absorptance, Reflectance, Transmittance, Optical damage threshold, Refractive index, Permittivity, Permeability, Skin depth
Mechanical properties	Density, Strength, Hardness, Ductility, Viscosity, Porosity, Poisson's ratio, Modulus of elasticity, Fracture toughness
Thermal properties	Thermal conductivity, Thermal expansion coefficient, Thermal shock resistance, Melting point, Boiling point, Specific heat capacity
Environmental sensitivities	Temperature, Humidity, Acidity/alkalinity, Water stability
Etc.	Electrical characteristics, Chemical characteristics, Machining characteristics

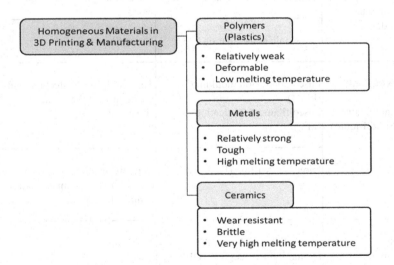

Fig. 3.2. Materials in 3D printing and manufacturing: polymers, metals and ceramics.

Therefore, their basic characteristics should be thoroughly understood and carefully considered. There are still other materials, e.g. paper, wood, stone and leather, which cannot be categorised into these three base materials; these materials are also used as base materials in special 3D printing and manufacturing methods.

Additive manufacturing processes are more sensitive to the form of base materials compared with traditional subtractive and formative ones [6]. Table 3.2 shows manufacturing technologies and their accessible base materials. Because of higher sensitivity of additive manufacturing to base materials, selection of the optimal laser and relevant sub-systems for the target material should be done with great care.

Table 3.2. Additive manufacturing technologies and accessible base materials.

	AM technologies	Materials
Laminated-based manufacturing (Solid)	Laminated object manufacturing	Metal film, plastic film, paper, wood
	Ultrasonic consolidation	Metal film
Extrusion-based manufacturing (Solid/Liquid)	Fused deposition moulding (FDM) Fused filament fabrication (FFF)	Thermoplastics, eutectic metals, metal clay, rubbers, clay, chocolate
	Direct ink writing (DIW)	Ceramic, ceramic matrix composite metal, metal matrix composite, cermet
Photo-polymerisation (Liquid)	Stereo-lithography (SLA) Digital light processing (DLP) Two-photon polymerisation (TPP)	Photopolymer

Table 3.2. (*Continued*)

Powder-based manufacturing (Powder)	Selective laser sintering (SLS)	Thermoplastics, metal powders, ceramic powders
	Selective laser melting (SLM)	Ti alloys, Cobalt Chrome alloys, stainless steel, Al, Cu, W, Ag, Au and Pt
	Electron-beam melting (EBM)	Cobalt Chrome, Titanium alloys, Stainless steel 316 recently added
	Selective heat sintering (SHS)	Thermoplastic powder
	Powder bed & inkjet head 3D printing	Almost any metal alloy & powdered thermoplastics

3.3. Polymers

A polymer is a large organic molecule composed of multiple repeated subunits. Both synthetic and natural polymers are nowadays playing important roles in 3D printing and manufacturing [10–13]. Polymers are created through the polymerisation of a large number of small molecules, known as monomers. The large molecular mass of polymers enables their unique mechanical properties such as toughness and viscoelasticity. Polymers linked by covalent chemical bonds have long been researched, and nowadays more research interests are on natural polymers.

3.3.1. *Plastic*

Plastic is a polymer that includes a wide range of synthetic or semi-synthetic organics, which can be moulded into solid objects with arbitrary shapes [14–16]. They are relatively cheap, easy to manufacture, impervious to water and have a long lifetime. These are the reasons why they have displaced traditional solid materials, such as wood, stone, metal, glass and ceramic. About 30% of plastic is used for packaging, another 30% for piping in building construction and another 20% for automobiles. Plastics can be categorised into two types: thermoplastics and thermosetting polymers (thermosets).

Thermoplastic

Thermoplastics are the plastics that do not experience chemical structure changes during the heating process. Thus, they can be repeatedly moulded into arbitrary shapes simply by heating. Most thermoplastics have a high molecular weight. The polymer chains are rapidly weakened with a minor temperature increase, which yields a viscous liquid status. As a result, thermoplastics have been regarded as proper materials in manufacturing processes for injection moulding, compression moulding and extrusion. Representative examples are polyethylene, polypropylene, polystyrene and polyvinyl chloride. The molecular weights of common thermoplastics lie between 20,000 to 500,000 amu (atomic mass unit), while thermosets have infinite molecular weights. Other examples in 3D printing are acrylic, acrylonitrile butadiene styrene (ABS), nylon, polylactic acid (PLA), teflon, polybenzimidazole, polycarbonate, polystyrene, polypropylene, polyether sulfone, polyetherether ketone, polyetherimide, polyethylene, polyphenylene oxide, polyphenylene sulfide and polyvinyl chloride.

Thermosets

Thermosets are the plastics that form irreversible chemical bonds during the curing process. Therefore, during heating process they do not melt; instead they decompose. When being cooled, they will not reform as well. The curing process can be induced by heating, chemical reaction or light irradiation. Thermosets are usually in the liquid or malleable state before the curing process and designed to be moulded into their final form. Examples are polyester resins, epoxy resins, vinyl ester and phenolic resins.

3.3.2. *Photopolymer*

Photopolymers are the polymers that change their structural and chemical properties when exposed to ultraviolet or visible light with high photon energy [17–19]. They are widely used in laser-based 3D printing processes, e.g. stereolithography apparatus (SLA), digital light processing (DLP) and two-photon polymerisation. Photopolymers can be deformed by the photo-curing process, where oligomers are cross-linked to form a thermoset network polymer. Their property changes can be initiated either by inherently existing internal chromophores or external photo-initiators. Photo-curing processes can be started by triggering the external photo-initiators by laser illumination at ultraviolet (UV) wavelength having high photon energy. With the aid of fluorescent dyes, visible lasers can also be used for the photo-initiation. This photo-curing results in the transformation of a mixture of monomers, oligomers and photo-initiators into a hardened solid polymer. To attain the physical properties required in 3D printing and manufacturing, various monomers and oligomers have been developed. Compared with other processes, photo-curing provides highly-selective polymerisation within the small volume where the UV laser light is focused on. By scanning a laser focal spot (corresponds to a unit volume, so-called voxel in 3D printing) over a photopolymer layer, one can construct arbitrarily-shaped 2D polymer structures. By repeating this process layer by layer, arbitrarily-shaped 3D polymer structures can be constructed. Compared with thermally cured polymers, photopolymers provide high resolution, high efficiency and a high polymerisation rate without the requirement for volatile organic solvents.

3.3.3. *Polymers in 3D printing*

Plastics are the most widely used 3D printing materials. Representative examples are Acrylonitrile Butadiene Styrene (ABS), Polylactic Acid (PLA) and polyvinyl alcohol (PVA). ABS is a common material (called a 'lego' plastic) with high strength, durability, impact resistance and toughness; thus being appropriate for moving parts, automotive parts and electronic housing. PLA is a user-friendly 3D printing material with moderate strength, durability and impact resistance, well-suited for consumer products requiring high print speeds. PVA is a water-dissolvable supporting material for PLA. Detailed material properties of these will be covered in this chapter.

ABS (Acrylonitrile Butadiene Styrene)

ABS is the 3D printing material that being most widely used in applications such as automotive parts, electronic housing, home appliances, musical instruments, sports equipment, toys, pipes and protective gears. It is because ABS is relatively inexpensive and can be easily 3D-printed by heating the nozzle to a temperature of ~250 °C. ABS is a strong, durable, partially flexible and heat resistant polymer. In the fused deposition modelling (FDM) process, filament-type ABS, in various colours, can be selected as the base material. In comparison to PLA, ABS plastic is the less 'brittle' plastic. ABS can be post-processed with acetone to provide a glossy finish. For 3D printing ABS, a printing bed that is temperature controlled around ~110 °C is required.

There are some limitations of ABS that need to be considered. First, it is not a bio-degradable polymer; however, it can be recycled. Second, the high-temperature process generates harmful fumes. Third, its performance can degrade when exposed to sunlight for a long time. If the ABS is not used for a long time, it should be stored in an air-tight container. ABS tends to attract moisture from the air, which could affect the printer's performance.

PLA (Polylactic Acid)

PLA is a biodegradable 3D printing material derived from renewable raw materials like starch (e.g. corn starch, sugar cane, tapioca roots or potato starch). PLA can be used in the form of resin for SLA or DLP processes as well as in filament form for FDM. PLA is a flexible user-friendly 3D printing material whose strength, durability and impact resistance are lower than those of ABS, but, a variety of colours can be realised with PLA. PLA is the easiest material to handle in 3D printing. It's extrusion temperature is around 180 ~ 230°C which is lower than that of ABS. A temperature controlled print bed (around 50 ~ 60°C) will be beneficial to the quality of the printed object but not essential. PLA is thermally less contractive and thus easy to print a large part with. However, it is brittle, so it is susceptible to sharp collisions. Because of its low toxicity, the installation of a fume hood is not mandatory. PLA absorbs water molecules from the ambient air so it is more sensitive to water absorption than other plastics; by the exposure, it will become brittle and sometimes difficult to print. Therefore, careful storage of the raw material is required. PLA is regarded as the more ecologically-friendly and safe material compared with other petrochemical-based plastics. Thus, it can be used in food storage, tableware and garments, hygiene products and biodegradable medical implants, e.g. tissue screws, sutures, pins, roads and tacks; as its degradation takes between 0.5 to 2 years.

PVA (Polyvinyl Alcohol)

PVA is a special water-soluble plastic. It is generally used as a paper-adhesive, thickener, packaging film or mould release agent. In the field of 3D printing, PVA is used together with PLA as a supporting structure with dual or multiple-extruders. Complex 3D-printed parts with numerous and complex overhung design structures must be developed with appropriate supports. Otherwise, the printed 3D structure would be deformed or can collapse. The printed object can be stored in water until the PVA has fully dissolved, without the need for time-consuming post-processing. PVA's extrusion temperature is similar with that of PLA, which is 180 ~ 230°C. However, it is not that easy to handle, because the moisture in the ambient air deteriorates the filament quickly. Therefore, PVA needs to be stored in a sealed box or container together with a desiccant and may need to be dried before use.

Nylon

Nylon is an extremely strong, durable and versatile thermoplastic used in 3D printing. It is made of repeating units linked by peptide bonds and is a type of polyamide. Nylon was the first commercially successful synthetic thermoplastic polymer and can be mixed with a wide variety of additives to achieve many different property variations. It is basically flexible when it remains thin; however, with high inter-layer adhesion, it is capable of serving as a living hinge and other functional parts. In the form of a filament, nylon naturally looks bright white with a translucent surface; the colour can be changed by absorbing acid-based clothing dyes. Because nylon is very sensitive to moisture as well, careful storage in desiccant, a vacuum or at an elevated temperature is recommended.

Table 3.3. Mechanical properties of different polymers in 3D printing.

Material	Tensile strength	Toughness (IZOD notched impact)	Heat deflection temperature
ABS	33 MPa	106 J/m	95.6 °C
Nylon	48 MPa	200 J/m	97.2 °C
PLA	50 MPa	80 J/m	65.5 °C
PC	68 MPa	53 J/m	138 °C
PEI	81 MPa	41 J/m	213 °C

Table 3.4. Polymers used in industrial applications.

Application area	Polymer examples
Manufacturing	ABS-ESD7, PC, PPSF, ABSPlus, Nylon, PC-ABS
Defense	ABS-ESD7, PPSF, ULTEM 9085, Nylon, PC-ABS
Medical	ULTEM 1010, PC-ISO, ABS-M30i
Consumer products	RGD525, VeroClear, ASA, DigitalABS, ABSPlus, Tango

3.4. Metals

Metals are materials that typically provide high strength, high electrical conductivity and high thermal conductivity [20]. Metals can be manufactured flexibly by subtractive, formative and additive manufacturing processes. They are removed by subtractive manufacturing due to their high strength; they can be permanently deformed by hammering, pressing or drawing out of their original shape without breaking or cracking; metal powders can also be sintered or fused through a melting process. A large number of atoms in the periodic table, 91 of the 118 elements, are metals. Examples widely used in laser manufacturing include stainless steel, aluminium, copper, gold, silver, platinum, titanium, brass and bronze.

3.4.1. *Mechanical properties*

One of the most important mechanical properties of metals is ductility, which is strongly related with the plastic deformation [21,22]. Reversible elastic deformation can be described by Hooke's Law, where the mechanical stress is linearly proportional to the strain. If mechanical forces larger than the elastic limit are applied, an irreversible permanent deformation is made, which is known as a plastic deformation. The non-directional nature of the metallic bond also contributes to the ductility of most metallic solids. When the ionic bonding plane slides, the location change shifts ions of the same charge into close proximity, which results in the crystal cleavage. Such a shift cannot be observed in covalently bonded crystals where fracture and fragmentation occurs instead. This irreversible change in atomic arrangement may be induced by an applied force (or work). An applied force can be tensile, compressive, shear, bending or torsion force. Fracture mechanics is an important tool for revealing the physics of stress and strain, which is essential for measuring, understanding and improving the mechanical properties of materials. Ductility governs the elastic and plastic deformation, as shown in Fig. 3.3. When a metal is elongated in a elastic deformation region, strain-stress follows a linear relationship. Then, if strain keeps increasing, the stress reaches to elastic deformation limit—the yield strength, where non-linear plastic deformation starts to be observed. If strain increases further, the stress touches the maximum value—the ultimate tensile strength (UTS)—and starts to decrease. With further strain increase, the metal finally fractures; the strain corresponding to this fracture is called a fracture elongation. Mechanical properties of several metals are shown in Table 3.5.

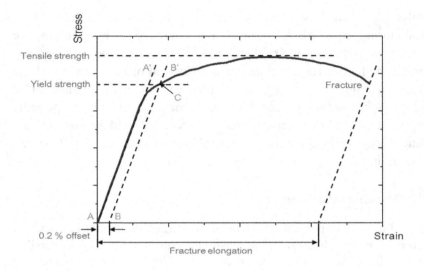

Fig. 3.3. Typical strain-stress curve of a metal. Ductility governs the elastic and plastic deformation. A-A': elastic deformation regime, B-B': 0.2% offset line from A-A' and C: yield strength.

Table 3.5. Mechanical properties of metals in 3D printing and manufacturing.

Material	Characteristics	Ultimate tensile strength	Yield strength	Elongation at break
Stainless steel 17-4 PH	• excellent weldability, • corrosion resistance • cost effective	980 MPa	500 MPa	25%
Stainless steel 316L	• excellent weldability, • corrosion resistance • ductility	640 MPa	530 MPa	40%
Aluminum AlSi10Mg	• low weight • high strength • good thermal properties	340 MPa	250 MPa	1.5%
Inconel 625	• high tensile strength • high creep strength • high rupture strength	900 MPa	615 MPa	42%
Titanium Ti64	• biocompatible • corrosion resistance	1,150 MPa	1,030 MPa	11%
Cobalt Chrome CoCrMo	• high tensile strength • good hardness • biocompatible	1,200 MPa	800 MPa	24%

Mechanical properties of 3D printed parts are strongly influenced by defects, thus product morphology such as pore volume fraction, pore size and powder spacing is important (see Table 3.6). 3D-printed metals generally provide high yield strength, low ductility and low dynamic properties as shown in Table 3.6 and Fig. 3.4. Fatigue limits in low-porosity materials are generally 0.35 UTS compared to 0.5 UTS for fully dense metals. Hot isotactic pressure eliminates internal defects and recovers the microstructure; this results in mechanical properties' improvement. Table 3.6 shows general trends in mechanical properties of 3D-printed metal and polymer parts.

Table 3.6. Mechanical properties of 3D printed materials (manufactured by SLS).

	Metals	Polymers	Non-Metallics
Modulus of Elasticity	Porosity Driven (Power Law)	Porosity Driven (Power Law)	Porosity Driven
Strength/Ductility	Porosity Driven Isotropic (High Δ)	Porosity Driven Anisotropic (Ductility)	Porosity Driven Weibull Works
Fatigue	$\sigma_e < 0.5$UTS or no σ_e		-
Fracture Toughness	Less or equal to bulk		-

3.4.2. Thermal and electrical properties

Metals have high electrical and thermal conductivity. They are typically malleable and ductile, deforming under stress without cleaving. In terms of optical properties, metals are shiny and lustrous. Sheets of metal beyond a few micrometres in thickness appear opaque, but gold leaf transmits green light. Although most metals have higher densities than most non-metals, there is wide variation in their densities; with lithium being the least dense solid element and osmium the densest. The alkali and alkaline earth metals in groups I A and II A are referred to as the light metals because they have low density, low hardness and low melting points. The high density of most metals is due to the tightly packed crystal lattice of the metallic structure. The strength of metallic

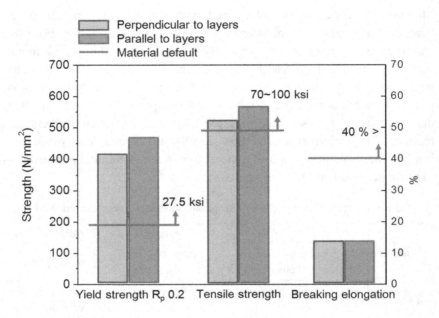

Fig. 3.4. Mechanical strength (yield strength, tensile strength and breaking elongation) of 316L stainless steel SLM, as processed.

bonds for different metals reaches a maximum around the centre of the transition metal series, as those elements have large amounts of delocalised electrons in tight binding type metallic bonds. However, other factors (such as atomic radius, nuclear charge, number of bonds orbitals, overlap of orbital energies and crystal form) are involved as well.

The electrical and thermal conductivities of metals originate from the fact that their outer electrons are delocalised. The electrical conductivity and heat conductivity of metals can be calculated from the free electron model, which does not usually take the detailed structure of the ion lattice into account. Atoms of metals readily lose outer shell electrons, resulting in a free flowing cloud of electrons within their otherwise solid arrangement. This provides the ability of metallic substances to easily transmit heat and electricity. Thermal properties of representative metals in 3D printing are shown in Table 3.6. In most metals, the maximum operating temperature is lower than 700 °C, thermal conductivity is ~15 W/m°C and thermal expansion coefficient is ~13 m/m°C.

Table 3.6. Thermal properties of metals in 3D printing.

Material	Maximum operating temperature	Thermal conductivity	Thermal expansion coefficient
Stainless steel 17-4 PH	550 °C	14 W/m°C	14×10^{-6} m/m°C
Stainless steel 316L	-	-	
Aluminium AlSi10Mg	-	-	
Inconel 625	650 °C	-	$12.5 \times 10^{-6} \sim$ 13.0×10^{-6} m/m°C
Titanium Ti64	350 °C	-	
Cobalt Chrome CoCrMo	1150 °C	13 W/m°C	$13.6 \times 10^{-6} \sim$ 15.1×10^{-6} m/m°C

3.4.3. *Metal alloys*

An alloy is a mixture of two or more material elements in which the main component is a metal. Most pure metals are too soft, brittle or chemically reactive so are not ideal for practical applications. Combining different kinds of metals to form alloys modifies the properties of pure metals so as to provide desirable characteristics. The general aim of making alloys is to make them harder, less brittle or corrosion-resistant. Iron alloys, i.e. steel, stainless steel, cast iron, tool steel and alloy steel, are the most widely used metal alloys by quantity and value. Iron alloyed with different relative ratios of carbon provides low-, mid- and high-carbon steels; increasing carbon levels reduces the ductility and toughness. While the addition of silicon makes cast irons, the addition of chromium, molybdenum and nickel to carbon steels (more than 10%) results in stainless steels. Other major metallic alloys are with aluminium, titanium, copper and magnesium. Copper alloys are most importantly used in electrical wiring. The alloys of aluminium, titanium and magnesium are valued for their high strength-to-weight ratios, which is therefore ideal for aerospace and special automotive. Alloys specially designed for highly demanding applications, such as jet engines, may contain more than ten elements.

3D printing can process various metals including titanium, nickel-base super alloys, stainless steels and tool steels, which are commercially available in the powder form. Printing materials include stainless steel 17-4PH, stainless steel 316L, aluminium AlSi10Mg, inconel 625, inconel 718, titanium Ti64 and cobalt chrome CoCrMo. These will typically perform with higher tensile strengths. Estimated prices of metal powders for 3D printing in 2012 are as follows (Wohlers Report):

- *Tool Steel*   *~$50/lb*
- *Stainless Steel*   *~$50/lb*
- *Aluminum Alloys*   *~$50/lb*
- *Co-Cr Alloys*   *~$55-250/lb*
- *Nickel Alloys*   *~$95-125/lb*
- *CP Titanium*   *~$150-400/lb*
- *Ti-6Al-4V*   *~$150-400/lb*

3.5. Ceramics

Ceramics are inorganic non-metallic materials made from compounds of a metal and a non-metal; they are a relatively new group of materials in laser-based 3D printing. Manufacturing of ceramics is based on heating and subsequent cooling. Clay was one of the earliest base materials used to produce ceramics, but various kinds of materials are used nowadays for industrial and building products. Ceramics are generally known to be strong, stiff, brittle, chemically inert and non-conductive (thermally and electrically) material. Crystalline ceramics cannot be processed with ease by a large range of methods. Available methods are categorised into two groups: one makes ceramics in the desired shape by *in-situ* reaction and the other forms ceramic powders into the desired shape, then solidifies the structure by sintering. The latter group includes methods like hand shaping, slip casting, tape casting, injection moulding, dry pressing and so on. Glass is not a ceramic because it is an amorphous and non-crystalline material; however, its processing steps and mechanical properties are similar with those of ceramics based on melting. In the molten state, glass can be shaped by casting or blowing into a mould.

3.5.1. *Mechanical properties*

Mechanical properties of ceramics are important in structural and building materials; the mechanical properties include elasticity, plasticity, tensile strength, compressive strength, shear strength, toughness, ductility and indentation hardness. As ceramics are based on ionic or covalent bonding, they tend to fracture before any plastic deformation, which is why ceramics have poor toughness. In addition, they are usually porous so the pores and other microscopic imperfections inside ceramics act as stress concentrators. Consequently, the toughness further decreases with the tensile stress. This results in catastrophic failures, as opposed to metals that have much more gentle failure modes. So as to overcome the brittle behaviour, ceramic development has introduced a relatively new class of ceramic matrix composites, in which ceramic fibres are embedded, so that fibre bridges may be formed across possible cracks. This process increases the fracture toughness of ceramics. The ceramic disc brake is a good example. Mechanical properties of representative ceramic materials are as shown in Table 3.7.

Table 3.7. Comparison of ceramic materials (with stainless steel).

Material	Density (g/cm^2)	Elastic modulus (GPa)	Flexural strength (GPa)	Fracture toughness $(MPa \cdot m^2)$	Max. service temperature (°C)
Aluminium oxide (sintered)	3.9	395	300	38	1,700
Zirconium oxide (sintered)	6.1	210	1,050	7	1,500
Silicon carbide (hot press)	3.1	400	380	3	1,600
Silicon nitride (reaction-bonded and sintered)	3.2	310	600	6	1,000
Boron nitride (hot press)	2.3	675	51	2.6	1,000
Silicon carbide (including fibre composite)	2.5	270	360	39	1,600
Advanced high-strength steel (QuesTek C61)	7.9	200	1,650	140	430

3.5.2. *Thermal properties*

The physical properties of a ceramic material fundamentally come from its chemical composition and crystalline structure; physical properties include odour, colour, volume, density, melting point, boiling point, heat capacity, physical form at room temperature (solid, liquid or gas state), hardness, porosity and refractive index. Most bulk-mechanical, thermal, electrical, magnetic and optical properties are significantly affected by the microstructure. The fabrication method and process conditions are generally determined by the microstructure because the reasons for many ceramic failures are from cleaved and polished microstructures. The microstructure includes grains, secondary phases, grain boundaries, pores, micro-cracks and structural defects.

Ceramics can withstand very high temperatures i.e. from 1,000 °C to 1,600 °C. An interesting feature of ceramic is its temperature-dependent conductivity. Temperature increase in ceramics can generate the grain boundaries, so some ceramic materials can suddenly become insulators (and act as a semiconducting material); these are mostly mixtures of heavy metal titanates. The semiconducting transition temperature can be designed over a wide range by chemical variations. In such ceramics, an electric current will be passed through until the material is heated up to the transition temperature, at which the circuit will be broken and the current will stop. Service temperatures and specific modulus of several ceramic materials are shown in Fig. 3.5.

3.5.3. *Electrical properties*

Regarding electrical conductivity, ceramic materials are widely used as electrical insulators. However, ceramic compounds can work as superconductors in some cases and some ceramics (such as zinc oxide) are semiconductors. While there are prospects for mass-producing blue LEDs from zinc oxide, ceramicists are more interested in the electrical properties which show grain boundary effects. Semiconducting ceramics

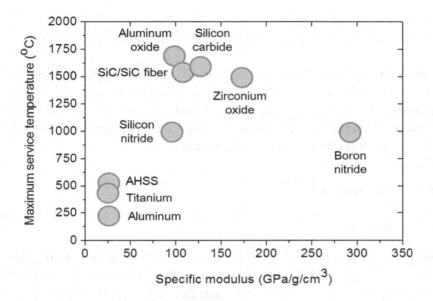

Fig. 3.5. Comparison of ceramic materials with metals.

can be utilised for gas sensors. When various gases pass over a polycrystalline ceramic, the electrical resistance will change; thereby, inexpensive gas sensing devices can be realised.

3.6. Composites

A composite material (also called a composite) is a material made from two or more constituent materials with significantly different physical or chemical properties that, when combined, produce a material with characteristics different from the individual components. The individual components remain separate and distinct within the finished structure. The new material may be preferred for many reasons: common examples include materials which are stronger, lighter, or less expensive compared to traditional materials.

Typical engineered composite materials include:

- Composite building materials, such as cements and concrete
- Reinforced plastics, such as fibre-reinforced polymer

- Metal composites
- Ceramic composites (composite ceramic and metal matrices)

Composite materials are generally used for buildings, bridges and structures such as boat hulls, swimming pool panels, race car bodies, shower stalls, bathtubs, storage tanks, imitation granite, cultured marble sinks and countertops. The most advanced examples perform routinely on spacecraft and aircraft in demanding environments.

3.7. Other Materials

3.7.1. *Papers*

Standard A4 paper is a 3D printing material utilised in laminated object manufacturing (LOM) or selective deposition lamination (SDL). In LOM and SDL, the paper is firstly unwound from a feed roll onto the paper stack and bonded to the previous layer using a heated roller. The roller melts a plastic coating or an adhesive layer on the lower side of the paper for the inter-layer bonding. Each layer is then patterned by laser cutting with the aid of XY galvano scanning mirrors. After the cutting process, the residual paper part is cut away to separate the layer. The extra paper is wound on a take-up roll. The advantages of using the paper as the base material are cost-effectiveness, safeness, reliability and full-colour printing capability. The 3D-printed parts made out of paper are also safe, eco-friendly and recyclable.

3.7.2. *Bio-materials*

There is a huge research trend moving towards 3D-printing bio-materials for fundamental studies to understand tissue/cell characteristics and in manufacturing functional human tissues for medical applications [23–34]. As an example, a living tissue is being investigated as a base 3D-printing material at leading institutions, for printing human organs for transplants, as well as external tissues for replacement body parts.

- Bio-ink: bio-ink comprises stem cells and cells from a patient. These can be laid down, layer by layer to form a tissue. Human organs such as blood vessels, bladders and kidney portions have been replicated using this technology.
- Bone material: a bone-like material comprising silicon, calcium phosphate and zinc was recently printed. This bone-like material was integrated with a section of undeveloped human bone cells. In a week, the growth of new bone was observed along with the structure. This material was dissolved and proved itself not to harm the patient.
- Skin: 3D printing of cells can also help skin regeneration. This could bring a change in the skin treatment method. The potential for regenerative medical application will be tremendous in wound healing, plastic surgery and cosmetics.

3.7.3. *Food printing*

Chocolate is an easy material for understanding the concept of 3D food printing. When in its molten state, chocolate has similar material characteristics with thermoplastics. Thus, various 3D printing methods based on extrusion or powder-sintering can be applied to chocolate. There are also 3D printers that work with sugar, pasta and meat. 3D food printing offers a range of potential benefits. It can be healthy and good for the environment because it can help to convert alternative ingredients such as proteins, from algae, beet leaves or insects into tasty products. It also enables food customisation; allowing the food produced to be adjusted to meet individual consumers' needs and preferences. Food design and decoration will definitely benefit from 3D food printing, as it allows the product to simultaneously please the eye and the mouth at the same time. Another intriguing prospect for 3D food printing is food printing in outer space. Currently, astronauts eat freeze-dried foods during their mission. In order to enhance their quality of living, NASA is now actively looking for ways to 3D print food in space; their first project is a 3D food printer for pizza.

3.7.4. *Carbon materials: graphene and carbon nanotubes*

Graphene and carbon nanotubes (CNTs) have received great attention for their promising potential applications in electronics, photonics, biomedical and energy storage devices, sensors and other cutting-edge technological fields, mainly because of their fascinating properties; such as an extremely high electron mobility, a good thermal conductivity and a high elasticity. Graphene is a single atomic layer of crystalline graphite and CNTs are cylindrical carbon structures with unique mechanical and electronic properties. A CNT can be thought of as a sheet of graphene (a hexagonal lattice of carbon) rolled into a cylinder. Graphene- or CNT-enhanced nano-composites provide superior characteristics compared to traditional plastics and other materials used in 3D printing. With the support of graphene and CNTs, printing materials can share the phenomenal properties of graphene and CNTs; they become mechanically strong, thermally conductive and electrically conductive while maintaining their flexibility and high transparency. 3D printing of graphene will enable the creation of various novel products, such as printed electronic circuits, sensors or micro-batteries.

3.8. Optical properties of materials in laser-based processing

Primary physical properties of engineering materials include electrical, thermal, magnetic and optical properties. Optical property is defined as the interaction between the material and an electromagnetic (EM) wave of light. Optical properties include absorption, reflection, transmission, refractive index, birefringence and so on. The electromagnetic spectrum spans a wide range from γ-rays ($\sim 10^{-12}$ m), to ultraviolet (UV), visible, near-infrared (NIR), infrared and radio waves ($\sim 10^5$ m) as shown in Fig. 3.6 [2]. In the competing theories of light, light can be considered as the waves having photon energies (E) of $E = h\nu = hc_0/\lambda$, where, h is the Plank constant, ν is the optical frequency, c_0 is the speed of light and λ is the wavelength of light.

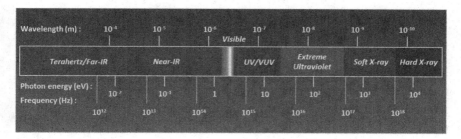

Fig. 3.6. Spectrum of EM waves in wavelength, frequency and photon energy domain.

3.8.1. *Wavelength dependent material absorption*

The interaction of photons with the electronic or crystal structure of a target material results in a number of physical phenomena. The photons may give their energy to the material (absorption); or photons of identical energy may be directly emitted by the material (reflection); alternatively, photons may propagate without any interaction with the material (transmission). At any moment of interaction, the total intensity of the incident light striking a surface (I_0) is equal to the sum of the absorbed (I_A), reflected (I_R) and transmitted (I_T) intensities: $I_0 = I_A + I_R + I_T$, due to conservation of energy.

If the material is not perfectly transparent, the intensity in the material decreases exponentially with propagation distance. When considering a material of small thickness, δx, the intensity falls during the propagation of δx and is expressed by $\delta I = -\alpha \cdot \delta x \cdot I$, where α is the absorption coefficient (dimension of m^{-1}). By considering the limiting case where $\delta x \rightarrow 0$, the intensity decrease can be expressed in a derivative form, $dI/dx = -\alpha I$. The solution of this derivative equation is $I = I_0 \exp(-\alpha)$. Penetration depth in a material is defined as the depth where $I/I_0 = 1/e$. Some examples of penetrations depths are as follows:

Penetration depth of materials
- *Water: 320 mm*
- *Glass: 290 mm*

- *Graphite: 600 nm*
- *Gold: 150 nm*

The absorption, transmission and reflection mechanism is different in metallic and non-metallic materials. In metals, electrons govern photon-material interaction, whereas in other materials the mechanism is dominated by the nuclei.

Absorption in metals

In metals, there exist partially-filled higher energy conduction bands. Thus, the incident photon energy on metals is used to excite the electrons to unoccupied states, which is why metals are opaque. Because of the high concentration of electrons, most photons are absorbed within 100 nm of the metallic structure. If the metal films are thinner than 100 nm, they will still transmit the light. After absorption, the excited atoms in the surface layers of metal atoms will relax, thus emitting the photons back at the same energy level, which is called a reflection. As a result, most metals reflect about 95% of the incident light. The remaining energy in the material will be converted into heat, which can be utilised for laser-based 3D printing. Metals are generally less reflective (more absorptive and transmittive) at high photon energy radiation (x-ray, γ-ray and ultra-violet) because the inertia of the electrons limits the reflection mechanism.

Material absorptions of several metals, aluminium (Al), silver (Ag), gold (Au), copper (Cu), molybdenum (Mo) and iron (Fe) are shown in Fig. 3.7. Metals have relatively high reflectivity at longer wavelengths than 800 nm. Therefore, Al, Ag and Au are widely used as broadband metal mirrors; Al provides broad reflection spectrum from ultraviolet to infrared but its reflectivity is relatively lower than others; Ag and Au provide higher reflectivity than Al; Ag is preferable mirror material for the visible wavelength, while Au is better in near-infrared wavelength. Iron provides higher absorption than 10% at wavelengths shorter than ~3.0 μm, which is the reason why iron looks less shiny than other metals.

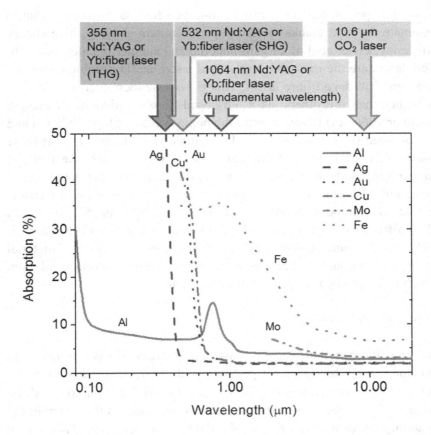

Fig. 3.7. Absorption of different metals (Al, Ag, Au, Cu, Mo and Fe) as a function of wavelength. Four representative wavelength regimes for laser-based 3D printing and manufacturing. X-axis is in logarithmic scale for showing wider wavelength range.

There have been three representative laser wavelengths in laser-based 3D printing; 10.6 μm, 1064 nm and less than 355 nm. As shown in Fig. 3.7, 1064 nm and ultraviolet wavelength can be solution candidates for laser processing of metals. Although all metals have high absorptance at the ultraviolet regime, excimer lasers in ultraviolet are too expensive, their maintenance is not easy and the beam mode is not good; therefore, they cannot be widely adopted in laser-based 3D printing. Instead, at 1064 nm, high-power lasers such as Nd:YAG, Nd:YVO₄, Yb-glass and Yb-fibre lasers are widely used because they are available in the market at reasonable prices. Recently, rare-earth-doped fibre lasers have begun to dominate the field thanks to their superior system stability, high energy efficiency, easier cooling of gain-material and simpler maintenance. Yb-fibre lasers are the most frequently used lasers in 3D printing compared to Er and Tm doped fibre lasers. This is because their energy efficiency is higher than the others, as a Yb-doped-fibre's emission wavelength (1030 or 1060 nm) is very near to the pump wavelength of ~975 nm and a Yb-doped-fibre laser is basically a three-level laser. If short-pulse lasers are used instead of continuous-wave (CW) lasers, the material temperature can be increased within a shorter time. Another advantage of using pulsed lasers is that non-linear wavelength conversion to visible {second harmonic generation (SHG) to 515, 530, or 532 nm} and ultraviolet wavelength ranges {third harmonic generation (THG) to 343, 353, or 355 nm} become possible; this can increase the material absorption rate and enable potentially higher manufacturing resolution, which is limited by the optical diffraction limit.

Absorption in non-metals

Non-metallic materials, such as plastics and ceramics, have energy band structures with bandgaps. If photons with energies greater than the bandgap (E_g) are incident on non-metallic material, the photons will be absorbed by giving their energy to electron-hole pairs in the material; the material may or may not re-emit the light when they relax. This is why most non-metallic materials have higher absorption at shorter wavelengths. Photon–material interaction in non-metallic material includes absorption, reflection and transmission; however, there is an additional important optical property, refraction. When photons are transmitted through a material, they cause electronic polarisation, which results in the reduced speed of the light in the medium. This also results in a change in the beam's direction when the photons are incident on the interface between two optical media. The refractive index (n) of a material governs refraction and reflection. If there is an interface between two transitive materials, one material with the refractive index of n_1 and the other material with the refractive index of n_2, the reflectivity, R at the interface is expressed as $R = \{(n_1 - n_2)/(n_1 + n_2)\}^2$.

The reflectivity of glasses is about 0.05 (5%). The refractive indices of several materials are shown in Table 3.8.

Table 3.8. Refractive index of a number of materials

Material	Refractive index	Material	Refractive index
Air	1.00	Silicon	3.29
Ice	1.31	Gallium arsenide	3.35
Water	1.33	Germanium	4.00
Silica glass	1.46	Teflon	1.35
Silicate glass	1.50	Polymethyl methacrylate	1.49
Sapphire, Al_2O_3	1.76	Polyethylene	1.52
Spinel, $MgAl_2O_3$	1.72	Epoxy	1.58
Rutile, TiO2	2.68	Polystyrene	1.60
Diamond	2.42	Salt, NaCl	1.54

Absorption in polymers

Generally, there are several kinds of polymers that are non-transparent, semi-transparent and transparent. Most of polymers strongly absorb the light in the ultraviolet and infrared wavelengths. Carbon-based optical-grade polymers are widely used for light-weight plastic optics. They are highly transparent at visible wavelengths; however, they also absorb the ultraviolet and infrared wavelengths. Therefore, for laser-based 3D printing of thermal processing, ultraviolet or infrared lasers should be used. This high-level absorptance of plastic at ultraviolet and infrared wavelengths is not readily apparent from the published data because such data is attained by using a very thin sample for the purpose of identifying the chemical structure. Figure 3.8 shows the optical absorptance of several polymer materials, PEEK, PPS and epoxy. PPS and epoxy are semi-transparent polymers so their transmittance at the visible wavelength between 500 and 1000 nm, is about 80%. Meanwhile, PEEK is a non-transparent polymer whose transmittance is less than 20%. All three polymers have about 100% absorptance at the wavelength below 250 nm. The absorptance at the infrared wavelength range, longer than 3000 nm, is also about 100%. Therefore, CO_2 laser at the 10.6 µm infrared wavelength have long been used for meting polymers for selective laser sintering or plastic cutting; ultraviolet excimer lasers can be utilised for this purpose as well, but they have not been widely used due to the high system cost. However, the use of ultraviolet lasers is a prerequisite in stereo-lithography apparatus (SLA) since high photon energy is required for the photo-polymerisation process.

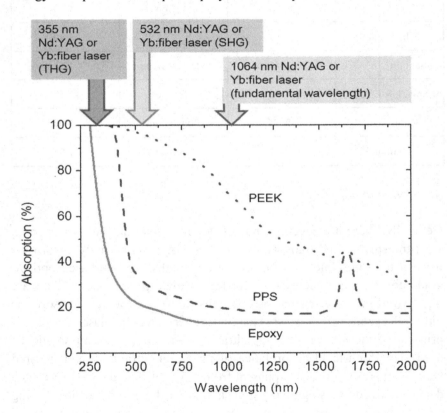

Fig. 3.8. Absorption of different polymers (PEEK, PPS and epoxy) as a function of wavelength.

Regarding the refractive index, carbon-based polymers have different properties compared to glasses or inorganic crystals. Refractive index is governed by the Abbe number of the polymer as shown in Fig. 3.9. Generally, the refractive index ranges from 1.30 to 1.73 in polymers (see Table 3.6). However, the refractive index is not homogeneous or uniform inside the 3D printed or moulded parts due to their different melt flow grades. Additives to regulate lubricity and colour can also produce subtle changes in the transmission properties.

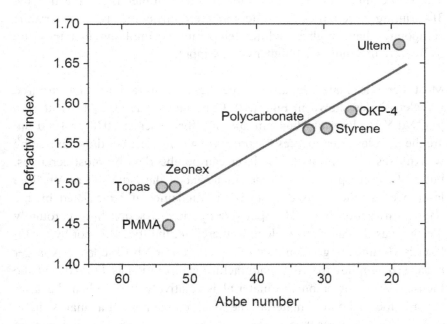

Fig. 3.9. Relationship between refractive index and Abbe numbers for widely used optical thermoplastics.

Absorption in ceramics

Laser sintering of ceramic materials in selective laser sintering (SLS) is basically similar to the polymer or metal laser sintering process, whereas ceramic sintering is based on a transient liquid-phase sintering. Ceramic SLS have achieved direct sintering of ceramics without polymer binder materials, but the absence of binder elements makes the laser-sintered ceramic part fragile and viable to breakage. Due to the complex interaction between the photons and ceramic materials, laser sintering parameters must be carefully selected to achieve desirable sintered parts for converting the photon energy efficiently to thermal heat. The melting point of a ceramic material is determined by its composition, powder granules shape, size, fluidity and homogeneity. With the aid of melting pools, the sintering process can be accelerated to a high laser scanning speed. In comparison with the established SLS of metals and plastics, the 3D printing of ceramics encounters inherent difficulties because ceramic compounds have higher melting temperatures, implying that they are difficult to melt and need higher energy input.

Most ceramic materials absorb the light in ultraviolet and infrared wavelengths as shown in Fig. 3.10. Continuous wave CO_2 laser at 10.6 µm, Nd:YAG laser at 1064 nm and Yb-fibre laser at 1030 or 1060 nm are the popular laser sources for processing ceramics, but they have some weaknesses. The infrared CO_2 laser can be absorbed by most ceramics, but its focused spot size is larger than that of the Nd:YAG or Yb-fibre laser because the focused spot size is fundamentally restricted by the optical diffraction limit. The spot size by the diffraction limit is roughly the same as the laser's wavelength; therefore, the focused spot by a CO_2 laser is 10 times larger than that of a Nd:YAG or Yb-fibre laser. A larger beam size implies that the manufacturing resolution is limited and the focused laser power on the material is relatively low. When the laser beam is focused onto a material, the local temperature at a small volume rapidly rises and reaches a very high value. The temperature can be higher than 3000 ^{0}C, which is over the 2000 ^{0}C required for special furnaces in traditional ceramic sintering. Moreover, the surrounding ceramic powders will not be significantly affected because the energy is highly localised within a small volume and the laser–material interaction time is much shorter than traditional sintering. This special feature can be achieved only by laser sintering; making it particularly suitable for SLS of ceramics with very high melting points.

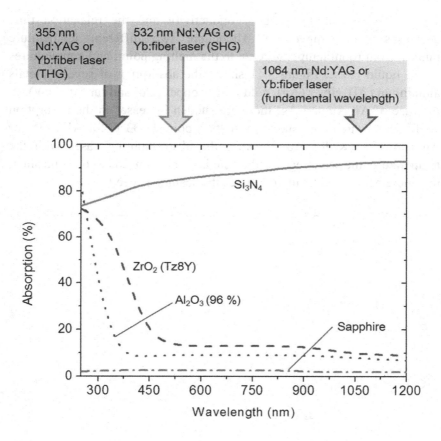

Fig. 3.10. Absorption of different ceramics (Si_3N_4, ZrO_2, Al_2O_3 and sapphire) as a function of wavelength.

3.8.2. *Temperature dependent material absorption*

Material absorptivity varies with the temperature. Therefore, knowledge on the absorption variation is of practical importance in understanding laser-based 3D printing. The temperature dependence of the optical properties of metals has been theoretically studied in many articles. By the Drude model of electrical conduction, it was shown that absorptivity increases with temperature when inter-band contributions are negligible.

This is because both the DC conductivity and the relaxation time decreases with temperature. There are two different temperature ranges: from room temperature up to the melting point; and temperatures in the liquid phase. Figure 3.11 shows the absorption of several metals aluminium (Al), silver (Ag), gold (Au), copper (Cu) and tungsten (W). In the case of Al, Au and Ag, there are sudden increases in the absorption coefficient, which correspond to melting points; 933 K for Al, 1235 for Ag and 1337 K for Au. Because the absorption increases with the temperature, the scanning speed of the laser beam needs to be controlled; this increment could be up to 5 times the scanning speed.

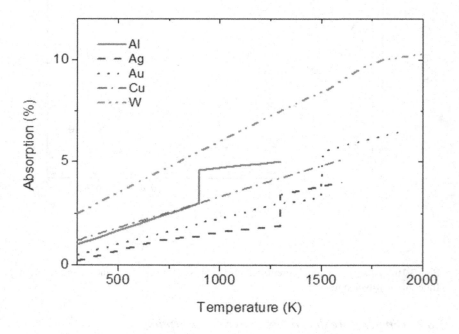

Fig. 3.11. Temperature dependent absorption of metals (Al, Ag, Au, Cu and W).

3.8.3. *Polarisation dependent material absorption*

Laser light is an electromagnetic wave having both electric and magnetic components in it. The relative orientation of these components with the beam's propagation direction is called polarisation as shown in Fig. 3.12.

The electric and magnetic fields' vector orientations are perpendicular to each other. The electric vector of the laser beam contains the power for laser material processing. Therefore, the electric vector's orientation and its temporal stability are the key parameters in processing metals and a number of dielectric materials. Figure 3.13 shows iron's (Fe) material absorption dependence on the polarisation and incidence angle with a 532 nm laser beam at room temperature. At the normal incidence (when incidence angle is 0), the absorption rate is the same for p- and s-polarisations. However, at an incidence angle of around 80 degrees, the difference in absorption rate is about 70%; 10% for an s-polarised beam; and 80% for a p-polarised beam.

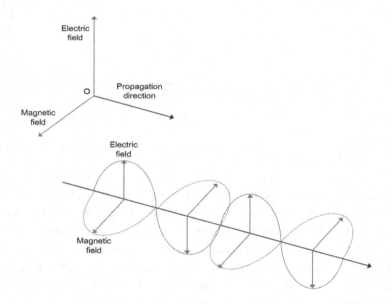

Fig. 3.12. Orientation of electric and magnetic field vectors.

A linearly polarised beam has a fixed electric field direction whereas the electric field of a circularly or elliptically polarised laser beam changes its electric field orientation with time and position (see Fig. 3.14). Time- and position-dependent relative phase delay between s- and p-polarisation parts makes the polarisation orientation rotate in circular and

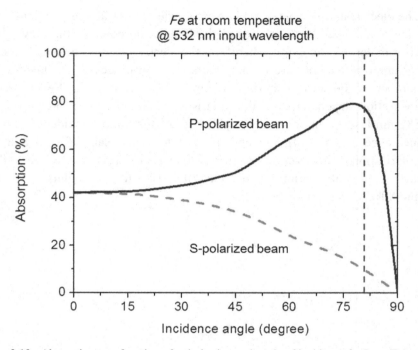

Fig. 3.13. Absorption as a function of polarisation and angle of incidence for Iron (Fe) at 532 nm. Note the maximum absorptance, A_p is at angle θ_B, the Brewster angle.

elliptical polarisation. White-light sources such as sunlight, halogen lamps or LED light radiate randomly polarised light that do not have a specific polarization direction, lasers generally emit linearly polarised light; for this purpose, there are polarisation selection components in laser oscillators, such as, the Brewster window, the polarising beam splitter and different types of polarisers. The polarisation state of a linearly polarised beam can be converted into either a circularly or elliptically polarised state, by using birefringent optical components. Birefringent crystal has different refractive indices for two polarisation components, so the relative phase between s- and p-polarisations can be changed. By precisely controlling the birefringent crystal's thickness and direction, one can rotate, convert or restore the polarisation states; a half-wave plate (HWP) rotates the electric field's orientation of a linearly polarised light; a quarter-wave plate (QWP) converts the linear polarisation into a circular one or reverts the circular polarised light into a linear one; the combination of HWP and QWP can be used for synthesising arbitrary polarisation states.

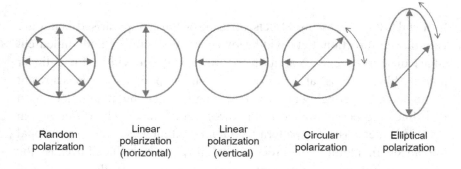

| Random polarization | Linear polarization (horizontal) | Linear polarization (vertical) | Circular polarization | Elliptical polarization |

Fig. 3.14. Polarisation states of the light: random, linear, circular and elliptical polarisation.

When a laser beam is circularly or randomly polarised, the electric field that contains the laser power is evenly orientated in all directions. Therefore, no polarisation sensitive effect can be seen in 3D printing and manufacturing. However, a linear polarisation means that the laser power is concentrated in one fixed polarisation direction. Therefore, laser-based manufacturing in line, at an angle, and perpendicular to the polarisation direction causes variations in 3D printing and manufacturing results.

When cutting transparent polymer materials, the polarisation effect is strongly dependent on the material's thickness. There are two extreme examples, described as follows. Firstly, if the polymer's thickness is over a certain threshold and a linearly polarised beam is focused onto the material, the strain-dependent birefringence in the polymer deteriorates the linear polarisation degree, so there will not be a significant polarisation effect. Secondly, if the thickness is over a certain threshold and a circularly polarised beam is incident on the polymer material, the top surface will be uniformly processed because the polarisation state is circular; however, after passing through a certain thickness, the strain-induced birefringence converts the circular input polarisation into an elliptical one, having relatively strong polarisation along one axis, which results in a non-uniform laser machining with polarisation sensitivity. Therefore, polarisation and thickness need to be carefully considered in transparent polymers.

In metals, the processing effects of using a linearly polarised beam are variation in directional cut-width, lower process stability and angled cut edges on curved cuts. Variation in orthogonal cut widths for a linear polarised beam (illustrated in Fig. 3.15) can be up to 30%. This is a result of the beam power being absorbed differently when the input polarisation is in-line, and perpendicular to the direction of travel. The difference in cut widths decreases the pattern resolution and increases the dimensional tolerances. In addition, the process's stability can be reduced because the change in cut widths interrupts stable processing. If the laser-based processing is done only along one direction, a linearly polarised laser beam along the processing direction is the fastest solution; however, in two- or three-dimensional laser processing, circular or random polarisation would be the simpler and better solution for uniformity.

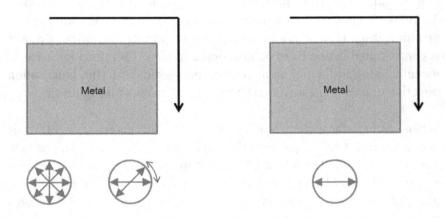

Fig. 3.15. Polarisation effect in the cut width. Direction of linear polarisation indicated in circles.

In laser-based manufacturing of ceramics, the dielectric nature of ceramics produces polarisation sensitivity. The result of this polarisation sensitivity is the difference in manufacturing depth when processing in-line and perpendicular to the polarisation direction. The scribe depth is the maximum when the polarisation axis is in parallel to the scribe direction; whereas, the cutting depth is decreased when the polarisation axis is in perpendicular to the scribe direction. A circularly polarised beam provides a uniformly processed manufacturing quality in all directions but at a lower processing speed.

Problems

1. Why lasers are regarded as the most important energy source for 3D printing nowadays? List four main reasons.

2. How small a spot can we focus the light into when a Yb-fibre laser of 1060 nm wavelength is used as the light source? What is the fundamental limitation behind this focusing capability?

3. How can one categorise 3D printing by its base material? List three examples per category.

4. List five critical questions that can be raised in the material selection process. Provide three answers per question.

5. Describe what kind of material properties need to be considered in laser-based 3D printing. Sort those properties into four categories.

6. What are the three base homogeneous materials in 3D printing?

7. What are the differences between the thermoplastics and thermosets used in laser-based 3D printing? List three examples of each type.

8. Which kinds of 3D printing processes use photopolymers as the base material?

9. Compare the mechanical properties (i.e. tensile strength, toughness and heat deflection temperature) of ABS, PLA, PVA and nylon.

10. Draw a stress-strain curve which can visualise the elastic and plastic deformation characteristics of a metal. Explain what these parameters indicate in the stress-strain curve, yield strength, tensile strength, fracture and elongation.

11. Compare the maximum service temperatures (°C) and specific modulus (GPa/g/cm^3) of ceramics and metals. What are the pros and cons of using ceramics compared with using metals?

12. Suggest three laser systems for 3D printing using metals and explain the reasons for your selections. What could be the ideal laser system for 3D printing if one can attain any wavelength, output power and pulse duration in the future?

13. Explain the refractive index of transparent (polymer and ceramic) materials that governs transmittance, reflectance and refraction at the interface between two materials.

14. Suggest three laser systems for the 3D printing of polymers and explain the reasons for your selections.

15. Suggest two laser systems for selective laser sintering (SLS) of ceramics and explain the reasons for your selections.

References

[1]   Yariv, A. (1976). *Introduction to optical electronics*.
[2]   Saleh, B. E., Teich, M. C., & Saleh, B. E. (1991). *Fundamentals of photonics* (Vol. 22). New York: Wiley.
[3]   Paschotta, R. (2008). *Encyclopedia of laser physics and technology*. Berlin: Wiley-vch.
[4]   Chryssolouris, E. L. K. E. (2013). *Laser machining: theory and practice*. Springer Science & Business Media.

[5]   Dubey, A. K., & Yadava, V. (2008). *Laser beam machining—a review*. International Journal of Machine Tools and Manufacture, **48**(6), 609–628.

[6]   Chua, C. K., & Leong, K. F. (2015). 3D *printing and additive manufacturing: principles and applications: principles and applications*. World Scientific, Singapore.

[7]   Thomas, D. S., & Gilbert, S. W. (2014). *Costs and cost effectiveness of additive manufacturing*. US Department of Commerce. Consulted at: http://nvlpubs. nist. gov/nistpubs/Special Publications/NIST. SP, 11, 76.

[8]   Kruth, J. P., Wang, X., Laoui, T., & Froyen, L. (2003). *Lasers and materials in selective laser sintering*. Assembly Automation, **23**(4), 357–371.

[9]   Khoo, Z. X., Teoh, J. E. M., Liu, Y., Chua, C. K., Yang, S., An, J., ... & Yeong, W. Y. (2015). *3D printing of smart materials: A review on recent progresses in 4D printing*. Virtual and Physical Prototyping, **10**(3), 103–122.

[10]   Lipson, H., & Kurman, M. (2013). *Fabricated: The new world of 3D printing*. John Wiley & Sons.

[11]   Gibson, I., Rosen, D., & Stucker, B. (2014). *Additive manufacturing technologies: 3D printing, rapid prototyping, and direct digital manufacturing*. Springer.

[12]   Tymrak, B. M., Kreiger, M., & Pearce, J. M. (2014). *Mechanical properties of components fabricated with open-source 3-D printers under realistic environmental conditions*. Materials & Design, **58**, 242–246.

[13]   Gebhardt, A. (2007). *Rapid Prototyping–Rapid Tooling–Rapid Manufacturing*. Carl Hanser, München.

[14]   Ahn, S. H., Montero, M., Odell, D., Roundy, S., & Wright, P. K. (2002). *Anisotropic material properties of fused deposition modeling ABS*. Rapid Prototyping Journal, **8**(4), 248–257.

[15]   Leigh, S. J., Bradley, R. J., Purssell, C. P., Billson, D. R., & Hutchins, D. A. (2012). *A simple, low-cost conductive composite material for 3D printing of electronic sensors*. PloS one, **7**(11), e49365.

[16]   Bose, S., Vahabzadeh, S., & Bandyopadhyay, A. (2013). *Bone tissue engineering using 3D printing*. Materials Today, **16**(12), 496–504.

[17]   Wilson, J. E. (1974). *Radiation Chemistry of Monomers, Polymers, and Plastics*. Marcel Dekker, NY.

[18]   Lawson, K. (1994). UV/EB Curing in North America, *Proceedings of the International UV/EB Processing Conference*, Florida, USA, May 1⁺5, 1.

[19]   Reiser, A. (1989) *Photosensitive Polymers*. John Wiley, NY.

[20]   Fung, Y. C. (1965). *Foundations of solid mechanics*. Prentice Hall.

[21]   Kruth, J. P., Leu, M. C., & Nakagawa, T. (1998). *Progress in additive manufacturing and rapid prototyping*. CIRP Annals-Manufacturing Technology, **47**(2), 525–540.

[22]   Rack, H. J., & Kalish, D. (1974). *The strength, fracture toughness, and low cycle fatigue behavior of 17-4 PH stainless steel. Metallurgical Transactions*, **5**(7), 1595–1605.

[23]   Bak, D. (2003). *Rapid prototyping or rapid production? 3D printing processes move industry towards the latter*. Assembly Automation, **23**(4), 340–345.

[24]   McGurk, M., Amis, A. A., Potamianos, P., & Goodger, N. M. (1997). *Rapid prototyping techniques for anatomical modelling in medicine.* Annals of the Royal College of Surgeons of England, **79**(3), 169.

[25]   Yang, S., Leong, K. F., Du, Z., & Chua, C. K. (2001). *The design of scaffolds for use in tissue engineering. Part I. Traditional factors.* Tissue engineering, **7**(6), 679–689.

[26]   Yang, S., Leong, K. F., Du, Z., & Chua, C. K. (2002). *The design of scaffolds for use in tissue engineering. Part II. Rapid prototyping techniques.* Tissue engineering, **8**(1), 1–11.

[27]   Leong, K. F., Cheah, C. M., & Chua, C. K. (2003). *Solid freeform fabrication of three-dimensional scaffolds for engineering replacement tissues and organs*. Biomaterials, **24**(13), 2363–2378.

[28]   Yeong, W. Y., Chua, C. K., Leong, K. F., & Chandrasekaran, M. (2004). *Rapid prototyping in tissue engineering: challenges and potential*. Trends in biotechnology, **22**(12), 643–652.

[29]   Leong, K. F., Chua, C. K., Sudarmadji, N., & Yeong, W. Y. (2008). *Engineering functionally graded tissue engineering scaffolds*. Journal of the mechanical behavior of biomedical materials, **1**(2), 140–152.

[30]   Cheah, C. M., Chua, C. K., Leong, K. F., & Chua, S. W. (2003). *Development of a tissue engineering scaffold structure library for rapid prototyping. Part 1: investigation and classification*. The International Journal of Advanced Manufacturing Technology, **21**(4), 291–301.

[31]   Lipton, J., Arnold, D., Nigl, F., Lopez, N., Cohen, D. L., Norén, N., & Lipson, H. (2010, August). Multi-material food printing with complex internal structure suitable for conventional post-processing. *In Solid Freeform Fabrication Symposium* (pp. 809–815).

[32]   Ivanova, O., Williams, C., & Campbell, T. (2013). *Additive manufacturing (AM) and nanotechnology: promises and challenges.* Rapid Prototyping Journal, **19**(5), 353–364.

[33]   Zhu, C., Han, T. Y. J., Duoss, E. B., Golobic, A. M., Kuntz, J. D., Spadaccini, C. M., & Worsley, M. A. (2015). *Highly compressible 3D periodic graphene aerogel microlattices.* Nature Communications, **6**, 6962.

[34]   Compton, B. G., & Lewis, J. A. (2014). *3D-printing of lightweight cellular composites.* Advanced Materials, **26**(34), 5930–5935.

Chapter 4

ONE-, TWO-, AND THREE-DIMENSIONAL LASER-ASSISTED MANUFACTURING

In this chapter, the fundamentals of laser-assisted manufacturing are discussed with illustrative examples. Both continuous wave and short pulse laser-assisted manufacturing are illustrated; showing the differences in their material removal mechanisms. The chapter begins with an introduction on laser machining, highlighting the advantages it can offer compared to conventional machining processes. A detailed account on one-, two- and three- dimensional laser machining processes are discussed. The chapter ends with a discussion on the various laser material processes and the role of focusing the laser beam for the purpose of achieving the desired effects.

4.1 Introduction

Laser beam machining (LBM) is a non-traditional manufacturing process, in which a **laser** is directed towards the workpiece for machining or material processing [1–4]. This process uses the supplied energy (thermal energy when using IR lasers) or pulses' energy (accumulated fluence when using short or ultra-short pulse lasers) to remove material from metallic or non-metallic surfaces. The laser is focused onto the surface to be worked, and the energy of the laser is transferred to the surface, either for heating, melting, or vaporising the material, or for ablation; depending on the type of lasers employed [1,5,6]. Though traditionally used for machining brittle materials with low conductivity, the advent of new generation lasers in the recent decade has changed the whole landscape, and lasers can now be used on most materials, thus setting the platform for its use in 2D and 3D printing.

4.2 Laser Beam Machining (LBM)

Laser machining can be divided into one-, two- and three- dimensional processes by differentiating the kinematics of the erosion front during beam–material interaction (see Fig. 4.1). However, all these processes exhibit common characteristics such as molten layer formation, possible plasma formation and beam reflection from the erosion front [1,5,7].

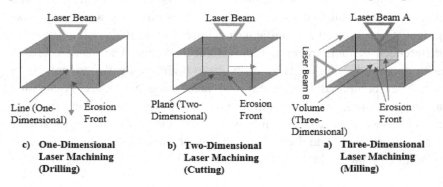

Fig 4.1. Drilling, Cutting and Milling using LBM.

Laser Drilling (One-dimensional LBM):
The laser beam is stationary relative to the workpiece. The erosion front located at the bottom of the drilled hole propagates in the direction of the line source in order to remove material [8].

Laser Cutting (Two- dimensional LBM):
The laser beam is in relative motion with the workpiece. Material removal occurs by moving the line source in a direction perpendicular to the line direction, thereby forming a two-dimensional surface. The erosion front is located at the leading edge of the line source [1,8].

Laser Milling (Three-dimensional LBM):
Two or more laser beams are used, thereby each beam providing a surface through the relative motion with the workpiece. The erosion front

for each surface is found at the leading edge of each laser beam. When the surfaces intersect, the three-dimensional volume bounded by the surfaces is removed [8].

Material removal rate is governed by the propagation speed of the erosion front. This is dependent on the erosion front's speed in the beam direction (in the case of 1D LBM). The scanning velocity determines the rate at which the 2D surfaces increase in the workpiece (in the case of 2D LBM), and the speed at which the two 2D surfaces produced by two laser beams determine the time required to remove a particular volume of the material (in the case of 3D LBM).

Dimensional Accuracy is determined by (a) the hole taper for laser drilling (1D LBM), (b) Kerf geometry for laser cutting (2D LBM) and groove shape for laser milling (3D LBM). However, the ***Surface Quality*** for all the LBM processes is related to (a) surface roughness (b) dross formation, and (c) the heat affected zone (HAZ) [9,10].

4.2.1   *One- dimensional: laser drilling*

The laser beam is a directional heat source and hence can be viewed as a one-dimensional line source with a line thickness equal to the beam's diameter. In principle, this method is governed by an energy balance between the laser irradiation and the conduction of heat into the workpiece, the energy lost to the environment, and the energy required for a phase change in the workpiece. Figure 4.2 depicts this energy balance by showing the different energy loss mechanisms and energy required for machine or material processing.

Advantages of laser drilling are [11–13]:
  (a) Holes can be made on materials such as ceramics, hardened metals and composites which are normally difficult to machine with conventional machines.
  (b) Higher accuracy and smaller dimensions can be achieved; with the proper selection of laser beam parameters (power, pulse

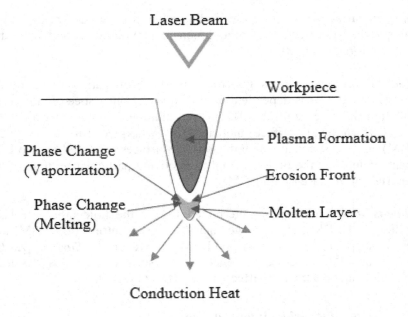

Fig. 4.2. One-dimensional machining—the process.

    rate, focusing lens and interaction time), the desired geometry can be achieved.

(c) High drilling rate can be achieved.
(d) Eliminate the need for tool changes.
(e) Laser allows holes to be drilled at high angles of incidence.

However, in many cases, unless the system has a good CNC-controlled automation arm, holes with stepped up diameters cannot be drilled.

4.2.2   *Two-dimensional: laser cutting*

In laser cutting, a kerf is formed through the relative motion between the laser beam and the workpiece surface (see Fig. 4.3), and the workpiece's thickness is equal to the depth of cut. The temperature inside the workpiece is dependent on the distance to the erosion front and is independent of time. The molten material accumulated at the erosion front is expelled using the coaxial gas jet.

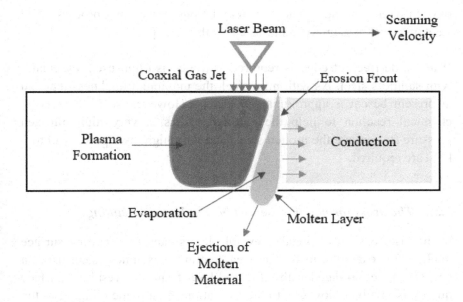

Fig. 4.3. Two-dimensional machining—the process.

The cutting depth of a laser is directly proportional to the quotient obtained by dividing the power of the laser beam by the product of the cutting velocity and the diameter of the laser beam spot [14], given by,

$$t \propto \frac{P}{vd},$$

where, t is the depth of cut, P is the laser beam power, v is the cutting velocity and d is the laser beam spot's diameter.

The depth of the cut is also influenced by the workpiece's material. The material's reflectivity, density, specific heating and melting point temperatures all contribute to the laser's ability to cut the workpiece.

4.3 Oxy-Laser Cutting:

To achieve laser cutting, lasers require a gas to assist in the cutting. Oxygen is used when cutting mild steel as the chemical reaction of oxygen with iron releases heat, thereby helping the cutting process. Mild steel of a thickness up to 25 mm is cut in this way [15].

If an oxide-free cut edge is required, nitrogen is often used, especially with stainless steel. As with mild steel, the maximum thickness that can at present be cut is approximately 25 mm. However, since there is no chemical reaction to help the cutting process, a very high nitrogen pressure (to remove the molten steel) and very high laser power (up to 7 kW) are required.

4.3.1 *The dross formation observed in stainless-steel cutting:*

In the laser cutting of metals, several factors can affect cutting surface quality. For example, in the laser-gas cutting of stainless steel using a coaxial gas jet arranged with a focused laser beam, the resulting surface quality is relatively low due to the formation of an oxide clinging to the bottom. This is known as a "dross" formation [16,17]. The dross is mainly composed of iron oxides and chromium oxides. Due to their higher melting point, the dross material re-solidifies before the gas jet can completely expel them from the kerf.

This dross formation can be reduced by the intelligent use of gas mixtures such as 60% CO_2 and 40% O_2, which reduces the formation of oxides and helps to eject the molten material. It also improves the weldability of the cutting kerf surface.

4.3.2 *Advantages and limitations:*

Advantages of laser cutting can be identified as given below:
- High material removal rate (workpiece thickness up to 10 mm).
- Kerf widths narrower than those achievable with mechanical means.
- Can be cut from the curved workpiece.
- No residue/debris while cutting paper/wood/composites.

Drawbacks of this process always lie in the following aspects:

- Efficiency decreases as workpiece thickness increases.
- Tapered kerf edge is observed (due to beam divergence) compared to the straight vertical kerf walls achievable by conventional methods. However, this can be subdued by focusing the laser beam on the interior of the workpiece rather than at the surface.

4.4 Three-dimensional processes

Three-dimensional processes such as laser grooving, laser scribing, and laser milling are all concerned with bulk material removal [1,18].

Figure 4.4 shows the way the laser beam impinges the material surface, how conduction heat is evolved, the ejection of molten material, plasma formation, and the use of the gas jet and related aspects. The optical configuration for this generally uses two intersecting laser beams to remove the desired volume of material.

Advantages:

- Three-dimensional laser machining can perform turning, threading and milling operations on materials which are difficult to machine mechanically due to high brittleness/hardness and abrasiveness.
- Scribing or marking can be used to make a permanent identification that can withstand more wear than those made through conventional means.
- Ideal for micro-machining—tremendous applications in the semiconductor industry.

Disadvantages:

- For metals and ceramics, the formation of molten material. Use of an off-axis jet can minimise the molten layer.
- Groove depth can fluctuate due to disturbances caused by laser beam changes, mechanical vibrations, material impurities and gas jet fluctuations. This unevenness in groove depth can decrease surface quality and the mechanical

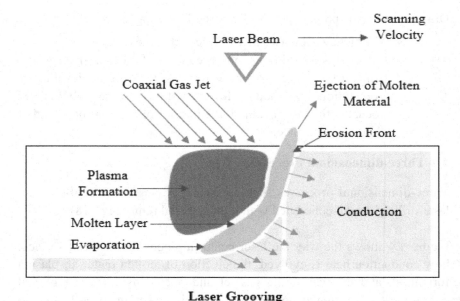

Laser Grooving

Fig. 4.4. Three-dimensional machining—the process.

strength of the finished part. A closed-loop control system can provide consistency in groove depth.

Further, by changing the angle between the beams used for the process, we can do all types of machining as detailed in Fig. 4.5. It explains the effects of the illumination angle of the laser beams on machining.

Figure 4.6 Explains the milling and turning by orienting the laser beams at 90^0 [12,17].

4.5 Basic Components of a LM System

A laser-based machining process requires the implementation and integration of a number of optical, electrical and mechanical components into a machining system. Basically, this includes four major subsystems such as laser beam generation, beam delivery, workpiece positioning and the related auxiliary devices as shown in Fig. 4.7 [1,12,13].

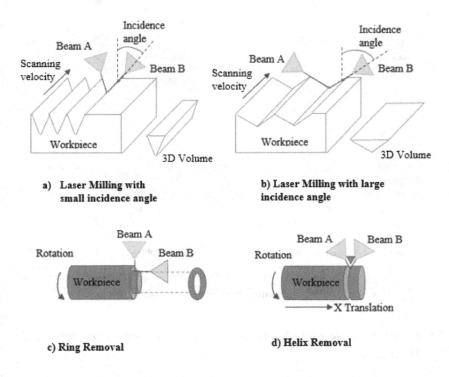

Fig. 4.5. Effect of incident laser beam angle on machining.

Laser beam generation:
This is accomplished by a laser device. CO_2 and Nd:YAG lasers are normally the most common lasers used for beam generation in industrial machining systems.

Beam Delivery:
This includes a combination of optical components that focus the beam onto the surface of the workpiece. Major components of the delivery system are given as follows:

(i) polarisers (ii) mirrors (iii) beam splitters (iv) focusing lenses and (v) fibre couplers/pigtailed fibres.

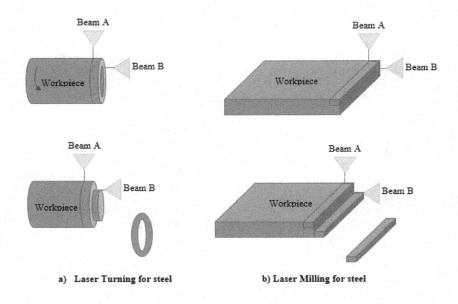

a) **Laser Turning for steel**   b) **Laser Milling for steel**

Fig. 4.6. Milling and turning by orienting the laser beams at 90^0.

4.6 Effect of State of Polarization (SOP) of Laser Beam on Machining

Photons propagate in space with an oscillatory motion that has oscillations in its electric field and thus produces an electromagnetic wave. This wave represents the direction and magnitude of the photon's electric field vector as a function of time. For a laser light beam, due to its highly coherent nature, the electric field for all photons in the laser beam are aligned in the same direction. This property leads to a linear polarisation of the laser beam. This property of the state of polarisation of the beam has a certain influence on the cutting and grooving operations using lasers [5,19,20]. By allowing to pass a linearly polarised beam through a quarter wave plate, a circularly polarised beam can be realised. Figure 4.8 shows line diagrams representing the linear and circular state of polarisations.

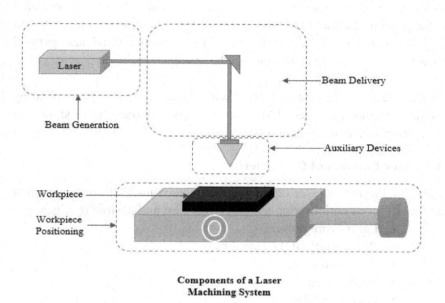

**Components of a Laser
Machining System**

Fig. 4.7. Basic components of an LBM.

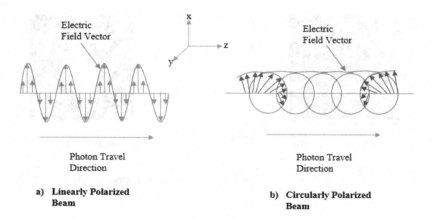

Fig. 4.8. Linear and Circular polarisation- Representative diagrams indicating the orientation of the respective electric field vectors.

When the electric field is oriented in the same direction as the scanning velocity, a *deep narrow cut* is obtained; and when it is oriented orthogonal to the beam scanning direction, a *wide shallow cut* is obtained [1]. When the angles differ by 5^0 or more from normal and parallel directions, a *groove curvature* would be visible.

These directional effects on cutting with a laser beam can be avoided by using a circular polariser [1,18]. A circularly polarised beam shows no preference in cutting direction (see Fig.4.9).

4.7 Laser Equipment Characteristics

Some of the major laser equipment characteristics that are detrimental and which need optimisation to achieve the best and desired results are:
- Laser beam power
- Wavelength of the laser beam
- Temporal mode
- Spatial mode
- Focal spot size

(a) Circular Polarisation   (b) Linear Polarisation

Fig. 4.9. Effect of polarisation on machined surface.

4.7.1 *Laser beam power*

If the power is below the required limit, it will result in an increased processing time or inability to perform machining operations on the desired materials. If the power is above the required limit, it increases the total expense.

The amount of required optical power is determined by examining the workpiece material's optical and thermal properties. As far as thermal properties are concerned, they can be divided into two categories: fixed thermal properties and loss properties. The amount of energy required to melt or vaporise the material falls under fixed properties. This depends on the heat capacity, latent heat and heat of vaporisation [21,22]. For ceramic materials, high power is required due to their high latent heat.
Energy transmitted to the surrounding material during processing belongs to the loss category. If it is a transient operation, thermal diffusivity is the dependent parameter; and if it is a steady state operation, thermal conductivity is the dependent parameter.

Lasers' optical properties affect the surface of the workpiece where the laser beam impinges. *The absorptivity of the material has the largest influence on the power requirements.* Absorptivity also depends on the wavelength, surface roughness, temperature, material's phase, and any use of surface coatings.

4.7.2 *Wavelength of the laser beam*

Wavelength is defined as the characteristic spectral length associated with one cycle of vibration for a photon in the laser beam. The absorptivity of the material depends on the wavelength. Hence, certain lasers are only suitable for processing certain materials [23]. E.g. aluminium and copper show low absorptivity to a 10.6μm CO_2 laser beam, and high absorptivity to a 1.06 μm Nd:YAG laser beam [24].

The absorption of the laser beam's energy depends on both the wavelength of the laser radiation and the spectral absorptivity characteristics of the materials processed. Copper and aluminium exhibit very high reflectivity to a CO_2 laser's radiation (10.6 micron wavelength), while a Nd:YAG laser's radiation has high absorptivity to these metals. Hence, it is more effective as the energy losses are minimal.

4.7.3 *Temporal mode*

The temporal mode can apply to both continuous wave (CW) or pulsed mode. CW mode has the advantage of producing smooth surfaces after machining, while also bearing the disadvantages of requiring high electrical power, and creating a heat affected zone (HAZ) in the process [1,25].

Pulsed mode operation of the beam has an edge in deep drilling or cutting applications. The disadvantages are the formation of wavy surfaces due to the periodic output, and if not properly controlled.

4.7.4 *Spatial mode*

Resonator design is critical for the production of a proper wavelength. The phase of an electro-magnetic wave may differ by resonator design, resulting in changes in the laser beam's spatial beam profile (say, the Gaussian beam TEM_{00}) [32]. Fig 4.10 shows the TEM_{00} Gaussian beam's profile.

Normally TEM_{00} is best suited for laser machining due to the following reasons:
- Phase front is uniform.
- Smooth drop of irradiance from the beam centre.
- Minimum diffraction effects.
- Allows generation of a small spot size.

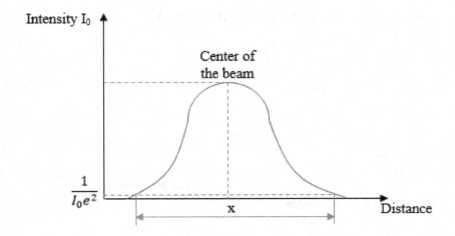

Fig. 4.10. TEM (Gaussian) profile.

4.7.5 *Focal spot size*

In material processing, the *irradiance* of the laser beam at the material surface is the important parameter. Irradiance is defined as power per unit surface area. Irradiances, great enough to melt or vaporise the material, can be obtained by properly focusing the laser beam. The maximum irradiance is usually at the lens focal plane, where the beam is at its smallest diameter. The location of this smallest diameter is called *focal spot*. Irradiances of the order of bW/cm^2 can be obtained at this spot. However, the imperfections of the optical components and diffraction effects limit the obtainable focal spot size [26,27].

There is a trade-off between a spot diameter and working distance (see Fig. 4.11.). It is noted from the relation between energy density and focal spot size that a decrease in spot diameter increases the energy density, but decreases the effective working range. The spot diameter can be decreased either by decreasing the focal length f or by increasing the unfocused beam diameter. Using a beam expander can increase the unfocused beam diameter [28]. If λ is the wavelength of the laser beam,

$$D_2 = \frac{8\lambda f}{\Pi D_1}, \text{ and } B = \frac{8\lambda}{\Pi}\left(\frac{f}{D_1}\right)^2$$

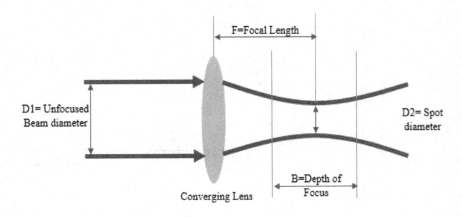

Fig. 4.11. Focal spot diameter and depth of focus—the trade-off.

4.7.6 *Factors affecting spot size*

- Incoming beam quality [beam divergence]
  Smaller divergence = very small focal spot size = High irradiance

- Diffraction
  When focusing a diffraction-limited laser beam with a lens, a higher f-number or larger focal length is required = larger focused diameter.

- Diameter of the incoming laser beam
  With a good beam anamorphic, or what is otherwise known as beam forming optics, a smaller focal spot size can be obtained by increasing the incoming beam's diameter.

Further, the heat losses are primarily due to conduction to the workpiece's interior and losses to the environment.

This may divert the beam's energy away from actual hole drilling process. This heat conduction depends on the material's thermal diffusivity α and interaction time t_i. The thermal penetration depth is given as [29],

$$\delta = \sqrt{\alpha \ t_i}$$

4.7.7 *Unique characteristics*

The unique characteristics of lasers for machining/material removal are considered to be the following [1,17]:

(A) ***Thermal processes:*** the efficiency of laser machining (LM) depends on the optical and—to a great extent—thermal properties of the material to be machined, rather than its mechanical properties. Hence, materials well-suited for LM should have, (i) high degrees of brittleness or hardness and (ii) low thermal conductivity and diffusivity.

(B) ***Non-contact process:*** energy transfer between the laser and the material occurs through irradiation. This has the following advantages:

    (i) No cutting forces are generated by the laser.
    (ii) No mechanically-induced damage on the material.
    (iii) No tool wear.
    (iv) No material vibration.
    (v) Due to the above, the material removal rate for LM is not limited by constraints such as maximum tool force, built-up edge formation, or tool chatter.

(C) ***Flexible process:*** by properly assembling with a multi-axis workpiece positioning system or a robot, the laser beam can be used for drilling, cutting, groove-welding and heat treating processes, using a single machine.

The advantages due to flexibility are:

(i) Elimination of the necessity of the transportation of specialised parts with a set of specialised machines.
(ii) Highly precise and small kerf widths or hole diameters compared to those achieved via mechanical techniques.

The disadvantages of LM (owing to being a thermal process) are:

(D) **Low energy efficiency**: This is due to the fact that in LM, the material removal occurs by melting or vaporising the entire volume to be removed. So the phase change of the material occurs atom by atom, which requires significantly higher energies and processing times than equivalent mechanical methods.

   However, this disadvantage has been overcome to a great extent by the high power industrial lasers now available; which offer high peak power and high quality beam forming optics. For example, a full-fledged 3-D LM provides an energy-efficient material removal method with a high degree of flexibility [1].

(E) **Material Damage**: During LM, high power densities are introduced on the workpiece's surface. In metals, the conduction heat resulting from the high energy density creates a heat affected zone (HAZ) in the vicinity of the erosion front [25] (see Fig. 4.12.).

In plastics and polymer matrix composites, material decomposition may occur as a result of elevated temperatures, which cause the breakdown of the polymer into charred residues and gaseous products. (*Also, there are issues related to poisonous gas*). Proper designing of the erosion front to remove the gaseous residues is an established technique now; hence this problem has also been solved to a great extent [25,30].

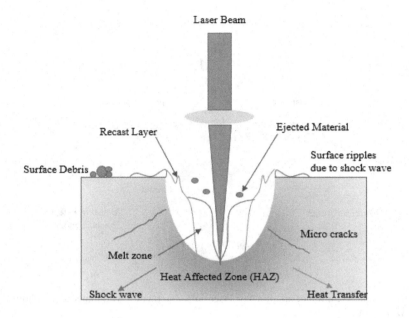

Fig. 4.12. Interaction of a CW laser beam with material—formation of HAZ and cracks.

4.8 CW Laser and UV Laser Machining

The physics of material removal for excimer lasers differs from that for CO_2 or Nd:YAG lasers. Instead of removing the material through melting or vaporisation, where the material is heated from solid to liquid and/or gaseous states, the excimer laser removes material through ablation, breaking the chemical bonds in it until the material dissociates into its chemical components (see Fig. 4.13.). When the bond breaking exceeds a critical value, the material decomposes. In this way, the pulse of the excimer laser beam removes material in pieces a fraction of a micrometre-thick, layer by layer [44].

In the case of CO_2 or ND:YAG lasers, the beam is focused onto a small area and traverses the workpiece to be machined. But excimer lasers produce large area beams that are masked through a template to achieve the desired cutting/machining area. After being passed through the mask,

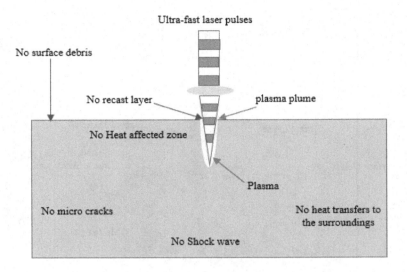

Fig. 4.13. Demonstration of an ultra-fast laser pulse interaction with material; and how no HAZ is created.

the beam is focused onto the workpiece through beam forming optics/a lens [12,31].

4.9 UV Laser Machining: Salient Features

Figure 4.14 shows an optical configuration for laser-assisted machining. Laser machining configuration with an Excimer laser follows the fundamental lens maker's formula: $1/u + 1/v = 1/f$, where

> u = distance from the mask to the lens
> v = distance from the lens to the workpiece
> f = focal length of the lens

The energy density of the laser beam on the workpiece is proportional to the square of the demagnification

$$ED_w = ED_o \, d^2$$

where

          d = demagnification factor = I u/ v I
          ED_w = energy density at the workpiece
          ED_o = energy density of the unfocused beam

Characteristics of the mask: Made out of metal that reflects most of the incident laser radiation so that the material underneath the mask is not removed during the cutting process [15].

Contact masking: Used when the workpiece area exceeds 1 mm x 1 mm and when there is no need for repetitive patterns to be cut. In this method, the mask is held close to the workpiece surface while the laser beam passes over it.

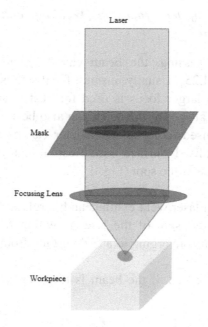

Fig. 4.14. Machining with Excimer Laser—optimal configuration.

Conformal masking: This is the other method of machining large areas. A thin metal layer of the mask is deposited on the workpiece in this method.

Mask projection has the following advantages,

- The mask is remote from the machining and hence does not suffer from debris damage.
- The use of a de-magnifying projection lens allows a reduction in the constraints on mask manufacturing.
- De-magnification means that the laser fluence on the mask is lower than on the sample, hence prolonging the mask's lifespan.
- The remoteness of the mask from the workpiece allows independent motion of the mask
- Multiple patterns can be superimposed on the sample by changing the masks.

4.9.1   *Optimization of lens focusing: Welding, cutting and heat treatment*

In laser material processing, the beam can be focused for different applications [10,13,14,25]. Usually, a small focal size is used for cutting and welding, while a larger focus is used for heat treatment or surface modification. The focal spot of the beam can also be varied based on the application. A defocused larger spot diameter can be used for surface heat treatment. Localised surface modification can be achieved by focusing on the surface of the spot (Fig. 4.15).

The cross-section of a laser weld exhibits an hourglass shape. This shape arises because the focal spot of the laser is within the material being welded, with the beam converging to and diverging from this spot.

For laser cutting, the focus of the beam is towards the bottom or back side of the plate.

The most common gas laser is the CO_2 laser, which emits light at a wavelength of 10.6 μm. Most metals better absorb the wavelength of light emitted by a Nd:YAG laser (1.06 μm) than that of a CO_2 laser. Also the Nd:YAG laser can be delivered via flexible fibre-optic cables, making it more versatile than the fixed delivery system of a gas laser.

4.9.2   *Factors Affecting the Choice between Gas and Solid-state Lasers:*

For CO₂ lasers: higher power, better beam quality in terms of focus ability, higher speeds and deeper penetration for materials that do not reflect its light, lower start-up and operation costs.

For Nd:YAG lasers: versatile fibre-optic delivery, easy beam alignment, easier maintenance, smaller equipment and more expensive safety measures than CO₂ because of its wavelength.

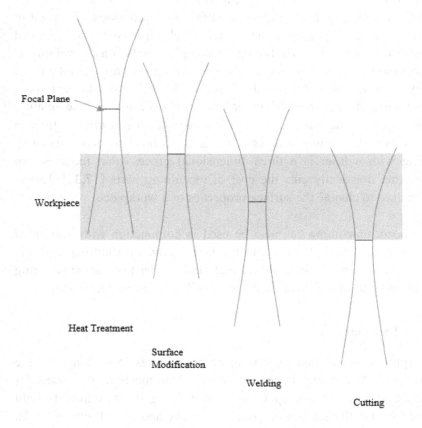

Fig. 4.15. Optimisation of lens focusing—for different material processing operations.

A typical Nd:YAG laser can be focused to a spot size of 0.1–0.8 mm (0.004–0.035 in), with power densities greater than 10^7 W/cm^2. At these power densities, a phenomenon referred to as **key-holing** occurs, which allows the continuous deep-penetration welding of metal. The metal melts and vaporises upon interaction with the beam; the pressure of the metal vapour pushes molten metal out of the way and forms a keyhole or cavity [32].

4.10   Laser Machining Applications

Lasers can be used for welding, cladding, marking, surface treatment, drilling, and cutting, among other manufacturing processes. It is used in the automobile, shipbuilding, aerospace, steel, electronics, and medical industries for precision machining of complex parts. Laser welding is advantageous in that it can weld at speeds of up to 100 mm/s, and has the ability to weld dissimilar metals. Laser cladding is used to coat weak parts with a harder material in order to improve the surface quality. Drilling and cutting with lasers are advantageous in that there is little to no wear on the cutting tool as there is no contact to cause damage. Milling with a laser is a three-dimensional process that requires two lasers, that drastically cuts the cost of machining parts [17,33]. Lasers can be used to change the surface properties of a workpiece.

Laser beam machining can also be used in conjunction with traditional machining methods. By focusing the laser ahead of a cutting tool, the material to be cut will be softened and made easier to remove; reducing cost of production and wear on the tool while increasing tool life[25].

4.11   Summary

The application of laser beam machining varies depending on the industry. In heavy manufacturing, laser beam machining is used for cladding and drilling, and spot and seam welding among others. In light manufacturing, the machine is used to engrave and to drill other metals. In the electronics industry, laser beam machining is used for wire stripping and the skiving of circuits. In the medical industry, it is used for cosmetic surgeries and hair removal [17].

Most of the lasers indicated in this chapter—such as CO_2, Nd:YAG, Nd:YVO4, Excimer and Femtosecond lasers—could find potential applications for 2D and 3D printing of structures and artefacts in the recent past. It is envisaged that new generation concepts and equipment based on these lasers, such as interferometric approaches, could produce high aspect ratio 3D features at the micro and nanoscale regime.

Problems

1. Explain using a simple diagrammatic sketch, how laser beam machining (LBM) can be divided into one-, two-, and three-dimensional processes.

2. What are the characteristics of laser machining equipment? Explain.

3. How does a focal spot's size determine the threshold power density (energy density) required in LBM?

4. Engineers using LBM always say that there is a definite trade-off between focal spot size and working range when using lasers for either machining or material processing applications. Give explanatory notes.

5. What is Oxy-Laser cutting and what are its significance and limitations?

6. Dross is formed while cutting certain materials while using LBM. What is dross and how can the formation of dross can be avoided or subdued?

7. Which parameters affect the thermal penetration depth while machining materials with laser beams?

8. By optimising the lens focusing in a LBM system, one can achieve cutting, drilling, heat treatment and surface modification. Explain this with a simple schematic diagram.

9. What is *Key-Holing*? Explain.

10. How does the state of polarisation of laser beams affect machined surface quality?

References

[1]   E. Chryssolouris, Laser machining: theory and practice, Springer Science & Business Media (2013).

[2]   J.K. Chua and V.M. Murukeshan, "UV laser-assisted multiple evanescent waves lithography for near-field nanopatterning," *IET Micro & Nano Letters* **4**(4), 210-214 (2009).

[3]   J. Bingi and V.M. Murukeshan, "Speckle lithography for fabricating Gaussian, quasi-random 2D structures and black silicon structures," *Scientific Reports* **5**, 18452 (2015).

[4]   R. Sidharthan and V.M. Murukeshan, "Pattern definition employing prism-based deep ultraviolet lithography," *IET Micro & Nano Letters* **6**(3), 109-112 (2011).

[5]   R. Menzel, Photonics: linear and nonlinear interactions of laser light and matter, Springer Science & Business Media (2013).

[6]   R. Sidharthan and V. Murukeshan, "Periodic feature patterning by lens based solid immersion multiple beam laser interference lithography," *Laser Physics Letters* **9**(9), 691 (2012).

[7]   Y. Deng, H. Zheng, V. Murukeshan, X. Wang, G. Lim and B. Ngoi, "In-Process Monitoring of Femtosecond Laser Material Processing," *International Journal of Nanoscience* **4**(04), 761-766 (2005).

[8]   H. Hocheng, Machining technology for composite materials: Principles and practice, (2011).

[9]   I.A. Choudhury and S. Shirley, "Laser cutting of polymeric materials: An experimental investigation," *Optics and Laser Technology* **42**(3), 503-508 (2010).

[10] N. Muhammad, D. Whitehead, A. Boor and L. Li, "Comparison of dry and wet fibre laser profile cutting of thin 316L stainless steel tubes for medical device applications," *Journal of Materials Processing Technology* **210**(15), 2261-2267 (2010).

[11] H. Huang, L.M. Yang and J. Liu, "Micro-hole drilling and cutting using femtosecond fiber laser," *Optical Engineering* **53**(5), (2014).

[12] A.K. Dubey and V. Yadava, "Laser beam machining—a review," *International Journal of Machine Tools and Manufacture* **48**(6), 609-628 (2008).

[13] E.K.W. Gan, H.Y. Zheng and G.C. Lim, "Laser drilling of micro-vias in PCB substrates," in *Electronics Packaging Technology Conference, 2000. (EPTC 2000). Proceedings of 3rd*, 321-326 (2000).

[14] M. Behringer, "High-power diode laser technology and characteristics" In: *High Power Diode Lasers*, Springer (2007).

[15] C. Paul, A. Kumar, P. Bhargava and L. Kukreja, "Laser-Assisted Manufacturing: Fundamentals, Current Scenario, and Future Applications" In: *Nontraditional Machining Processes*, Springer (2013).

[16] D. Teixidor, J. Ciurana and C.A. Rodriguez, "Dross formation and process parameters analysis of fibre laser cutting of stainless steel thin sheets," *The International Journal of Advanced Manufacturing Technology* **71**(9-12), 1611-1621 (2014).

[17] P. Parandoush and A. Hossain, "A review of modeling and simulation of laser beam machining," *International Journal of Machine Tools and Manufacture* **85**, 135-145 (2014).

[18] S. Mishra and V. Yadava, "Laser beam micromachining (LBMM)–a review," *Optics and lasers in engineering* **73**, 89-122 (2015).

[19] R. Buividas, M. Mikutis and S. Juodkazis, "Surface and bulk structuring of materials by ripples with long and short laser pulses: Recent advances," *Progress in Quantum Electronics* **38**(3), 119-156 (2014).

[20] D. Clarke and J.F. Grainger, Polarized Light and Optical Measurement: International Series of Monographs in Natural Philosophy, Elsevier (2013).

[21] B. Hu and G. Den Ouden, "Synergetic effects of hybrid laser/arc welding," *Science and Technology of Welding & Joining* **539-543**, 3872-3876 (2013).

[22]   T.I. Zohdi, "Rapid simulation of laser processing of discrete particulate materials," *Archives of Computational Methods in Engineering* **20**(4), 309-325 (2013).

[23]   A. Kaplan, "Laser absorptivity on wavy molten metal surfaces: Categorization of different metals and wavelengths," *Journal of Laser Applications* **26**(1), 012007 (2014).

[24]   O. Aubreton, F. Mériaudeau and F. Truchetet, "3D digitization methods based on laser excitation and active triangulation: a comparison," in **9896**, 98960I-98960I-8 (2016).

[25]   J. Meijer, "Laser beam machining (LBM), state of the art and new opportunities," *Journal of Materials Processing Technology* **149**(1), 2-17 (2004).

[26]   S. Quabis, R. Dorn, M. Eberler, O. Glöckl and G. Leuchs, "Focusing light to a tighter spot," *Optics Communications* **179**(1), 1-7 (2000).

[27]   G.M. Lerman and U. Levy, "Effect of radial polarization and apodization on spot size under tight focusing conditions," *Optics express* **16**(7), 4567-4581 (2008).

[28]   A. Shinde, S.M. Perinchery and M.V. Matham, "Fiber pixelated image database," *Optical Engineering* **55**(8), 083105-083105 (2016).

[29]   J. Kim and H. Ki, "Scaling law for penetration depth in laser welding," *Journal of Materials Processing Technology* **214**(12), 2908-2914 (2014).

[30]   Y. Jamil, R. Perveen, M. Ashraf, Q. Ali, M. Iqbal and M.R. Ahmad, "He–Ne laser-induced changes in germination, thermodynamic parameters, internal energy, enzyme activities and physiological attributes of wheat during germination and early growth," *Laser Physics Letters* **10**(4), 045606 (2013).

[31]   K. Dreisewerd, "The desorption process in MALDI," *Chemical reviews* **103**(2), 395-426 (2003).

[32]   M. Schmidt, A. Otto and C. Kägeler, "Analysis of YAG laser lap-welding of zinc coated steel sheets," *CIRP Annals-Manufacturing Technology* **57**(1), 213-216 (2008).

[33]   B. Lauwers, F. Klocke, A. Klink, E. Tekkaya, R. Neugebauer and D. McIntosh, "Productivity Improvement Through the Application of Hybrid Processes" In: *Advances in Production Technology*, Springer (2015).

Chapter 5

LASER-BASED 3D PRINTING

There are three fundamental manufacturing processes for 3D structures: subtractive, additive and formative. Conventional manufacturing processes have long been based on subtractive and formative ones. Additive manufacturing (also called 3D printing) was first demonstrated for rapid prototyping in the 1990s and is now expanding its territories to include the efficient production of complex multi-material structures over a wide range of manufacturing applications. Lasers can be applied to all of these subtractive, additive and formative processes as highly efficient energy sources. Based on the initial form of a material, all 3D printing system can be categorised into (1) liquid-based, (2) solid-based and powder-based ones; the basic principle, system design, laser characteristics and applications of laser-based 3D printing systems will be studied in this chapter based on this categorisation.

5.1. Introduction

Traditional manufacturing processes have long been based on subtractive or formative processes of bulk solid materials. In a subtractive process, one starts the manufacturing with a single block of solid material which is larger than the final size of the desired product. Some portions of the material are removed until the desired shape is reached; this subtractive manufacturing includes traditional mechanical milling, turning, drilling, sawing, cutting and grinding. Meanwhile, in a formative manufacturing process, mechanical forces are applied on a thin sheet-like material so as to form it into the desired shape; this formative manufacturing includes bending, forging, press forming, rolling, spinning, drawing and injection moulding. From the 1990s, the third manufacturing process, additive manufacturing, begun to be commercialised [1–3]. Due to the remarkable

121

growth of information technology (IT), computer-aided drafting was first introduced in the 1970s and became popular in the 1980s in 3D product design, modelling and manufacturing. The concept of rapid prototyping (one of the different names of additive manufacturing; See Table 5.1) was then proposed and became commercially available from the 1990s. Rapid prototyping has saved time and cost in product design processes extensively compared to physical prototyping, i.e. the traditional construction of mock-ups in the automobile, aircraft and aerospace industries [4–8]. Time saving is getting more critical nowadays in order to meet fashion-sensitive market trends. An additive process is the exact reverse of the subtractive process; the end product is much larger than the base material present at the start of the process. Materials are manipulated layer by layer so that they are successively combined to form the desired object. There are more than 30 kinds of additive manufacturing technologies and over 100 different additive manufacturing machine models available today; about 70% of them are based on lasers.

Table 5.1. Different terms used for additive manufacturing.

Keywords	Other terms
Additive	Additive manufacturing, Additive fabrication, Additive processes, Additive techniques, Additive digital manufacturing, Additive layer manufacturing
3D	3D printing, 3D modelling
Layer	Layer manufacturing
Freeform	Freeform fabrication, Solid freeform fabrication
Rapid	Rapid prototyping, Rapid manufacturing, Rapid technology, Rapid tooling
Direct, Digital	Direct digital manufacturing, Direct tooling, Digital fabrication

Laser-based manufacturing can be categorised into three processes, namely, subtractive, additive and formative processes (see Fig. 5.1) [9,10]. Representative laser-based subtractive processes are laser cutting, drilling, and piercing; representative laser-based formative processes are laser forming, bending, and welding; laser-based additive manufacturing processes include Stereolithography Apparatus (SLA), Selective Laser Sintering (SLS), Selective Laser Melting (SLM) and Laser Engineering

Net Shaping (LENS). There are some technologies located at the intersection of two processes. In Laminated Object Manufacturing (LOM), each layer is processed by laser cutting (subtractive manufacturing), but the final product is completed by stacking those multiple layers (additive manufacturing). In laser bending and welding, some part of the base material melts down (subtractive manufacturing), but two thin materials are connected to design a shape by using their respective melted part (formative manufacturing).

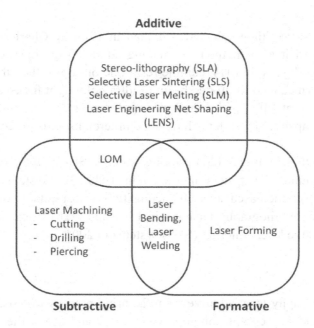

Fig. 5.1.   Laser-based manufacturing processes. These processes have been categorised by original and final specimen shape to subtractive-, formative-, and additive processes [1].

5.2. Liquid-Based 3D Printing

Most liquid-based 3D printing systems build the parts in a vat of photo-curable liquid resin. An organic resin solidifies under the exposure to light—usually in the ultraviolet (UV) range. The light, lamp, or laser beam cures the resin near the surface, forming a thin solidified layer.

Once the solidified layer of the part is formed, it is lowered by a piston-based elevation control system, to allow the next layer of resin to be coated and similarly form over it. This layer-by-layer printing process continues until the entire part is complete. The vat can then be drained and the part can be removed for further processing.

5.2.1. *Stereolithography Apparatus (SLA)*

History

The term "stereolithography" was coined in 1986 by Charles W. Hull, who patented it as a method for making solid objects by successively printing thin layers of a UV curable resin one on top of the other. Hull's patent described a concentrated beam of ultraviolet light focused onto the surface of a vat filled with liquid photopolymer. In 1986, Hull founded the first company, 3D Systems Inc., to commercialise this procedure.

Among all the commercial 3D printing systems, SLA is the pioneer with its first commercial system marketed in 1988. 3D Systems Inc. has grown through increased sales and acquisitions, most notably of the EOS GmbH's stereolithography business in 1997 and DTM Corp., the maker of the selective laser sintering (SLS) systems in 2001.

Principle

Stereolithography is an additive manufacturing process which employs a vat of liquid UV-curable photopolymer resin and scans the UV laser beam over to build parts' layers, one by one [11–18]. For each layer, the laser beam traces a cross-section of the part pattern on the surface of the liquid resin with the aid of Galvano scanning mirrors. Exposure to the UV laser light solidifies the pattern traced on the resin and joins it onto the layer below (see Fig. 5.2) [19–21]. The piston then lowers the elevation of the platform for the printing of the next layer. This elevation change determines the thickness of a single layer in the SLA, which is typically between 0.05 and 0.15 mm. Then, a blade sweeps over the printed part, to recoat the resin on top of the part. By repeating this process, completely designed 3D parts can be built. After the building process has been completed, the parts are immersed in a chemical bath to clean the excess resin out before being cured in a UV oven.

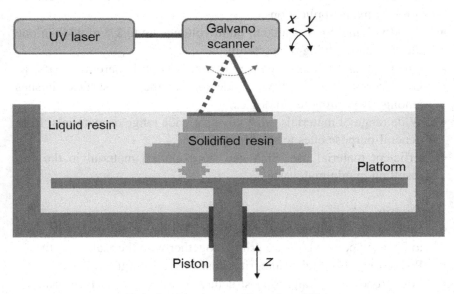

Fig. 5.2. The system configuration of SLA.

The processing speed is the most critical advantage of the SLA method. Fully functional 3D parts can be manufactured within a day; the time taken is dependent on the size and complexity of the part, and ranges from a few hours to a day. A large SLA machine can print a volume of $1500 \times 750 \times 550$ mm^3, which is good enough for printing automobile parts. Because the parts printed by SLA are strong enough, they can be used as master patterns for injection moulding, thermos forming, blow moulding, and metal casting. Although SLA resin has long been expensive (around \$200 per litre previously), it has recently gone down to \$50 per litre thanks to wider applications. In SLA, supporting structures are required to avoid gravity-induced vertical deflection, recoat-blade-induced lateral deflection and the unwanted separation of the parts from the elevation platform. These supports need to be manually removed from the finished parts.

The strengths and weaknesses of SLA are listed as below.

Strengths
- High accuracy. SLA has good end-to-end accuracy and thus can be used for many applications.
- Build volume. SLA is able to build volumes up to 1.5 m wide in one piece without gluing or assembly.
- Fast printing. SLA can create a full-sized car dashboard in two days.
- Good surface finish. SLA provides one of the best surface finishes among 3D printing technologies.
- Wide range of materials. SLA covers a wide range of materials; from general-purpose ones to specific ones.
- Efficient material use. SLA can store unused material in the vat, resulting in minimal waste.

Weaknesses
- Supporting structure. Structures with overhangs and undercuts need to have supports which are printed together with the main structure.
- Post-curing. For higher integrity, post curing is required.
- Post-processing. Supporting structure removal and surface grinding may be needed.

Lasers for SLA system

Because UV-sensitive photopolymers are usually used in SLA, the lasers for SLA must have emission wavelengths shorter than ~400 nm. As there are not many gain materials with a reasonable price and performance in this wavelength range, frequency-tripled NIR lasers are widely used. Crystal-based solid-state lasers (e.g. Nd:YAG and Nd:YVO$_4$) or Yb-fibre lasers at 1.06 μm centre wavelengths are the representative gain media. These gain materials are pumped by high-power flashlamps or 980 nm high power diode lasers. Both continuous-wave and pulsed lasers can be used for SLA itself but, pulsed operation is better for a stronger UV power. As the wavelength conversion from 1060 nm (NIR) to 353 nm (UV) (this process is called as frequency tripling or third harmonic generation (THG)) is a nonlinear optical process, the conversion efficiency increases in proportion to the input laser's peak power. Therefore, pulsed lasers are preferred for generating the UV wavelength as they usually provide a higher peak power than continuous-wave lasers. Wavelength conversion is made by focusing the laser beam into a nonlinear optical crystal, such as KTP ($KTiOPO_4$), LBO (LiB_3O_5) or BBO (beta-BaB_2O_4). Since wavelength conversion efficiency is very sensitive to the crystal's parameters, such as the crystal's cut-angle, walk-off angle, length, surface contamination and temperature, it needs to be air-sealed with temperature stabilisation. UV lasers are hazardous to both eye and skin; however, the power looks weaker than the actual power level because our eyes are not sensitive to UV. Therefore, proper safety instruction and precaution are prerequisites.

SLA machine example: 3D Systems' SLA system

3D Systems produce a wide range of AM machines to cater to various part sizes and throughput [22]. Current SLA models include the ProJet series and iPro™ series. The ProJet® 6000 and ProJet® 7000 series are low-cost SLA machines. ProJet printers have two sizes, three high definition print configurations and a wide range of VisiJet® SL print materials such as Tough, Flexible, Black, Clear, HiTemp, Impact and Jewel. For larger build envelopes, the iPro™ 8000, iPro™ 9000 and ProX™ 950 are available. These machines are used to create casting patterns, moulds, end-use parts and functional prototypes. The ProX™ 950 (see Figs. 5.3 and 5.4) has single-part durability for product builds up to 1.5 m wide in one piece without assembly [23]. It also provides high material efficiency because all unused material remains in the system for the next use, resulting in minimal waste. Specifications of these machines are summarised in Table 5.2.

5.2.2. *EnvisionTEC's DLP-based Perfactory system*

History

EnvisionTEC was founded with the name, Envision Technologies GmbH in 1999 and re-organised in 2002 by Mr Ali El Siblani. EnvisionTEC

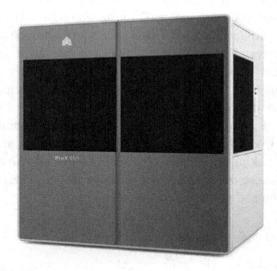

Fig. 5.3. 3D Systems' SLA 3D printing machine, ProX 950. (Courtesy of 3D Systems)

Table 5.2. Summary of specifications of the ProX 950.

3D Systems' ProX 950 Specifications	
PolyRay Print Technology	
✓ Lasers	Solid-state frequency-tripled Nd:YVO4 with SteadyPower
✓ Wavelength	354.7 nm
✓ Power (nominal) — at head	1,450 mW (1,000 mW at material surface under nominal optical path condition)
✓ Laser warranty	10,000 hours or 18 months (whichever comes first), replacement at 800 mW
Zephyr Recoating System	
✓ Process	Removable applicator
✓ Adjustment	Self-levelling; self-correcting
✓ Layer thickness	Min: 0.05 mm; Max: 0.15 mm

ProScan Scanning System	
✓ Border spot (diameter @ 1/e²)	0.13 mm
✓ Large hatch spot	Nominal 0.76 mm
✓ Maximum part drawing speed	
Border spot	3.5 m/sec (150 ips)
Large hatch spot	25 m/sec (1000 ips)
Build Envelope Capacity	
✓ MDM950 (ProX 950)	1500×750×550 mm³
✓ Maximum part weight	150 kg

Fig. 5.4. Parts printed by 3D Systems' ProX 950; (left) Engine block and (right) Automobile dash board over 1.5 m in length. (Courtesy of 3D Systems)

provides 3D printing solutions based on 2D-pattern-projection with higher productivity in wide applications. They also deal with software and material development to increase the systems' productivity and cost effectiveness. The company has two headquarters, one in Gladbeck, Germany and the other in Dearborn, Michigan, US.

Principle: 2D-pattern-projection DLP

DLP is a similar process to SLA in that it is a 3D printing process which works with photopolymers. The major differences are in the light source and beam delivery optics. DLP utilises conventional light sources, e.g. a UV lamp, and beam delivery optics, e.g. a liquid crystal display (LCD) or a deformable mirror device (DMD). Beam delivery optics transfers the 2D image pattern on the LCD or DMD to the entire surface of the vat of photopolymer resin (see Fig. 5.5) [24–28]. Compared to the SLA which requires 2D-scanning of the laser beam over the photopolymer vat, DLP can be faster because it is based on direct 2D projection without any mechanical beam scanning. As the key projection unit, a DLP projector based on the DMD is the most widely used. DMD is a display device based on MEMS technology that uses a small-sized digital micro-mirror array. It was developed in 1987 by Dr Larry Hornbeck from Texas Instruments. While the DMD was invented by Texas Instruments, the first DLP projector was developed by Digital Projection Ltd in 1997. DLP is used in a variety of display applications; from traditional displays to interactive displays and also non-traditional embedded applications, including medical, security and industrial uses. Similar to the SLA, DLP produces highly accurate parts with excellent resolution, but its similarities also include the same requirements for supporting structures and post-curing. One advantage of DLP over SLA is that a shallow vat of resin is required for the process, which results in less waste and lower running costs.

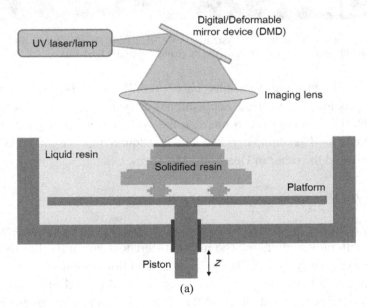

Fig. 5.5. The system configuration of DLP. (a) Top-down illumination DLP and (b) Bottom-up illumination DLP.

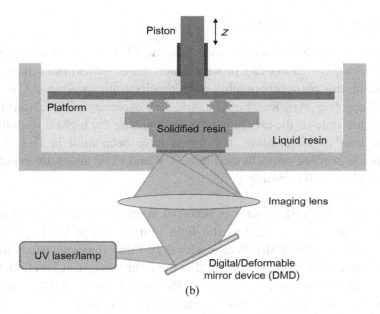

(b)

Fig. 5.5. (*Continued*)

The strengths and weaknesses of DLP are listed as below.

Strengths
- High building speed: 100 mm per hour (at 0.1 mm pixel height).
- No wiper or leveller required: DLP uses gravity for levelling.
- Small quantity of resin during the build.
- Office-friendly process.
- Less shrinkage.
- Safe supply cartridges.

Weaknesses
- Limited building volume: The 2D pattern on a DMD mirror can be projected onto the material within a limited field-of-view.
- Peeling of completed part.
- Requires post-processing.

Light source requirements

In DLP, UV lamps are widely used instead of UV lasers. Depending on applications, UV lasers can provide much better performance than UV lamps. In the photolithography of microelectronic products, UV and EUV lasers (e.g. Ar_2, Kr_2, F_2, Xe_2, ArF, KrF excimer or EUV lasers) are used in the state-of-the-art equipment for realising the highest resolution. Up to date, in 3D printing, UV lamps have been used in DLP for a reasonable printing price. In the future, UV and EUV lasers are expected to be applied to 3D printing.

In EnvisionTEC's 3D printer, a 100~180 Watt arc lamp based on solid tungsten electrodes is used as the light source. Due to its large beam divergence, an elliptical reflector with a 100 mm diameter needs to be incorporated with the UV lamp so that the light power can be efficiently transferred to the sample. When a UV laser is used in DLP, power transfer efficiency can be much better than a UV lamp (by more than hundreds of times) due to its high degree of coherence. With a 120 W lamp, 5200 lumens light is detected through a 6 mm aperture. Its average lifetime is 6000 hours at operation under 120 W.

DLP machine example: EnvisionTEC' Perfactory system

Perfactory is the AM system built by EnvisionTEC suitable for an office environment (see Fig. 5.6). In Perfactory, the photo-polymerisation is initiated by an image projection called DLP, which requires mask projection to cure the resin. The standard system achieves resolutions between 25 and 150 μm depending on the material. Additional options such as the 'Mini Multi Lens' system and the 'Enhanced Resolution Module' (ERM) enable designers to build smaller figures with high surface quality. The combination of the Perfactory 3, Mini Multi Lens, ERM and an 85 mm lens is able to create high-quality parts with a voxel size as small as 16 μm.

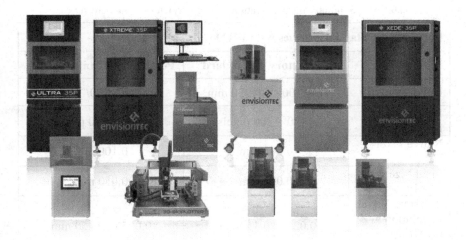

Fig. 5.6. EnvisionTEC's DLP and other 3D printer series. (Courtesy of EnvisionTEC)

'Mini Multi Lens' provides three choices of lenses with different focal lengths of 60, 75 and 85 mm, which enables the dynamic printing resolution in Z-axis from 15 to 150 μm. Optical diffraction limits the attainable resolution in 3D printing and manufacturing, which is determined by the lens aperture, focal length, light wavelength and the refractive index of the medium. Mini Multi Lens changes the focal length for a dynamic control of the resolution. To further improve surface finish and accuracy, most Perfactory systems are supplied with ERM. In ERM, there are two exposures for each voxel, laterally shifted by half a pixel, which halves the lateral (X- and Y-directional) resolution of the system. It is a very similar technique with sub-pixel resolving algorithms for optical microscopy. For example, a Perfactory with a resolution of 64 μm can provide a resolution of 32 um when using ERM. This provides an excellent surface finish, reduces pixelation effect and maintains accuracy to the intent of the 3D CAD design. Refer to Table 5.3 for Perfactory 4's specification summary.

Table 5.3. Summary of specifications of the Perfactory 4 standard series.

Perfactory 4 Standard Series with ERM		
	Perfactory 4 Standard	**Perfactory 4 Standard XL**
Build Envelope	160×100×180/230 mm³	192×120×180/230 mm³
Projector Resolution	1920×1200 pixels	1920×1200 pixels
Native Pixel Size	0.083 mm	0.100 mm
Pixel Size with ERM	0.042 mm	0.050 mm
Dynamic Voxel Resolution in Z (material dependent)	0.025 to 0.150 mm	0.025 to 0.150 mm
Data Handling	STL	

DLP printed parts and applications

ABS Tough (see Fig. 5.7(a)) is a tough 3D printing material for SLA; it is suitable for high-quality prototypes in automotive and consumer goods, as well as production-quality end-use parts. In dental applications, there are several 3D printing materials: E-Appliance for orthodontic models (see Fig. 5.7(b)), E-Dent for light-cured micro-hybrid filled dental crowns and bridges and E-guard for bio-compatible crystal clear splints and retainers. E-Dent is FDA approved. The E-Shell series has been specially designed for hearing aid industries; it is CE certified and Class-IIa biocompatible according to ISO 10993 (see Fig. 5.7(c)).

Fig. 5.7. DLP printed parts by (left) ABS Tough, (middle) E-guard and (right) E-shell. (Courtesy of EnvisionTEC)

5.2.3. *Liquid-based multi-photon micro-fabrication*

Two-photon polymerisation (TPP) is a 3D printing technique to fabricate 3D structures with resolutions down to 100 nm. When the laser beam from a femtosecond (10^{-15} s) pulse laser (emitting infrared light around 800 nm) is focused into a small spot in a photopolymer vat, polymerisation occurs only within the focal volume, where the intensity of the absorbed light is the highest. In contrast to previous SLA approaches, smaller features (smaller than the optical diffraction limit) can be manufactured without the need for supporting materials.

History

Nanoscribe GmbH offers 3D printers for the micro- and nanometre scale. Nanoscribe was founded in 2007 as a spin-off company from Karlsruhe Institute of Technology (KIT), where the technological basis was laid since 2001. The performance of their 3D printer, Photonic Professional GT system, was emphasised in February 2014 by being awarded a Prism Award in San Francisco, USA, the "Oscar of Photonics" in the "Advanced Manufacturing" category. Nanoscribe is located in Eggenstein-Leopoldshafen near Karlsruhe, Germany.

Principle: Two- or multi-photon polymerisation

In conventional SLA, a UV laser or lamp light with high photon energy is scanned over the surface of a photosensitive material, therefore producing 2D patterns of solidified photopolymer. The exposure to UV radiation (where a strong material absorption wavelength range is located) induces photopolymerisation through a single photon absorption process at the material surface. Therefore, with conventional SLA methods, it is possible to fabricate 3D structures based on the layer-by-layer printing approach only. As most photosensitive materials do not have absorption peaks in the infrared wavelength, they are transparent to infrared light. Therefore, one can initiate two-photon polymerisation (TPP) with short-pulse near-IR lasers within the small volume of the photosensitive material (without unwanted polymerisation in the volume nearby), which has not been possible in SLA with UV lights [29–35].

Figure 5.8 shows an illustration of the TPP-based 3D printing apparatus. Intense ultrashort laser NIR pulses are focused onto a small volume inside the UV-sensitive photopolymer; the small focused volume corresponds to a voxel in TPP-based 3D printing. In principle, multi-photon absorptions (for instance, three-photon absorption) can be used for photo-polymerisation for nano/micro 3D printing as well but they are not widespread to date because of their lower energy efficiencies compared to TPP. UV-sensitive photopolymer is polymerised along the 3D trace of the moving NIR pulsed laser focus, thus enabling the efficient manufacturing of designed 3D polymeric patterns. The photopolymerisation mechanism of UV-sensitive (λ_{UV}) polymer under the NIR light (λ_{IR}) is two-photon absorption, where the material absorbs two photons at a time, so the light wavelength is regarded to be half of the incident light ($\lambda_{IR}=2\lambda_{UV}$). As the TPP happens only when the focused peak intensity exceeds a certain polymerisation threshold level, very small 3D features (sub-micrometre even sub-100 nm in size) can be printed, which is smaller than the optical diffraction limit.

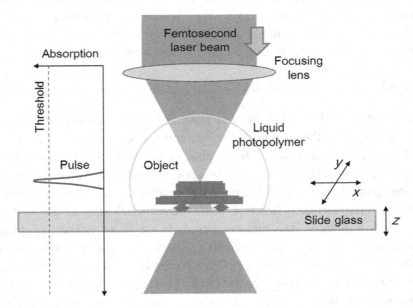

Fig. 5.8. The system configuration of nonlinear TPP-based 3D printing. (Courtesy of Nanoscribe GmbH)

The strengths and weaknesses of TPP are listed as below.

Strengths
- High resolution. The best printing resolution (less than a micrometre) can be realised with TPP.
- High accuracy. TPP provides good end-to-end accuracy. Depending on sample scanning mechanism, e.g. flexure-based 3D positioners with feedback sensors, nano-metre accuracy can be claimed.
- Best surface finish. TPP provides the best surface finishes among 3D printing technologies. Even optical quality surfaces can be printed.

Weaknesses
- Special laser requirement. Printed features are strongly dependent on the laser in TPP. For high-quality products, well designed high-peak-power femtosecond pulse lasers should be used.
- Slow printing. Because of high power requirement in TPP, the laser beam should be scanned over the volume in a point-by-point scanning manner, which makes the printing speed slow.
- Limited materials. TPP was first demonstrated in early 2000, so the number of available printing materials is still low.

Laser requirements for TPP system

Because TPP is based on nonlinear optical absorption, the high peak power of the laser pulse is the prerequisite. Therefore, utilising ultra-short pulse lasers, such as a femtosecond laser, is beneficial for the process. If the process is TPP, the second harmonic of the laser wavelength should lie in the absorption wavelength of photopolymers. For example, a Ti:Sapphire femtosecond laser light at a centre wavelength of 800 nm is effectively absorbed in the UV-curable resin at its half wavelength, 400 nm. As a compact, alignment-free and more stable alternative light source for a crystal-based Ti:Sapphire femtosecond laser, frequency-doubled Er-doped fibre femtosecond laser at 780 nm are widely used nowadays for TPP; note that the original wavelength of the Er-doped fibre femtosecond laser is 1560 nm. Using the two-photon absorption process, the UV-curable resin responds to the

390 nm wavelength; the second harmonic of the incident laser wavelength (780 nm). As another light source candidate, a Yb-doped fibre femtosecond laser provides the centre wavelength of 1060 nm so its second harmonic at 530 nm can be utilised for special kinds of resin with relatively longer absorption wavelengths covering 530 nm. Other possibilities of using a Yb-fibre femtosecond laser in photo polymerisation is either by using a multi-photon absorption process (where more numbers of photons participate in the process than in TPP); for example, by using its third harmonic wavelength of 353 nm or by using a second harmonic wavelength (530 nm) generated by an optical crystal as the excitation light source and using its TPP at 265 nm. Since both processes include more numbers of nonlinear wavelength conversion, the energy conversion efficiency for photopolymerisation is significantly lower than TPP.

In TPP, the two most important laser parameters are the wavelength and peak power; let us focus more on peak power here as we have already considered the wavelength in the previous paragraph. The peak power is a dependent parameter that is determined by four independent parameters: laser average power, repetition rate, pulse duration and focused spot size. The peak power is proportional to the average power but is inversely proportional to the other parameters of the repetition rate, the pulse duration and the spot size. In an Er-doped fibre femtosecond laser, the average power at a second harmonic wavelength of 780 nm is ~150 mW, the repetition rate (the number of pulses per second) is between 50 and 250 MHz and the pulse duration is ~100 fs. When focused onto a small spot of ~1.0 μm, the peak power is in the range of $10^9 \sim 10^{10}$ W/cm^2. If the material threshold for UV-curable resin is lower than this, the solidification process will be initiated.

Laser maintenance is a practical but important issue for broader applications in 3D printing. A Ti:Sapphire crystal-based femtosecond laser (at an 800 nm centre wavelength) is pumped by a solid-state Nd:YVO$_4$ continuous-wave laser (at a 532 nm centre wavelength which is the second harmonic of its fundamental wavelength at 1064 nm), which is in turn pumped by high power diode lasers (at 830 nm). This serial pumping process limits the wall-plug efficiency of the crystal laser to a lower level. There are two main maintenance concerns in crystal-based femtosecond pulse lasers; one is the troublesome laser cavity alignment process and the other is the lifespan of the pump diode lasers. As a crystal-based femtosecond laser's cavity comprises many bulk optics (e.g. mirrors, lenses, prisms and gratings) distributed over a large optical table of several metres in length, long-term environmental changes easily deteriorate stable pulsed operation; thus, daily or weekly realignment is required [36,37]. The warranty of the pump diode lasers (at 830 nm) is two years or 2500 hours' diode usage, whichever applies first. Compared to crystal-based lasers, rare-earth-doped fibre femtosecond lasers are much better for maintenance. They are basically alignment-free because the entire cavity is made of optical fibres which are fusion-spliced with one another. Moreover, rare-earth-doped fibres such as Er and Yb are pumped by 980 nm diode lasers that are well-established for optical communication; the lifetime of the pump lasers can reach 25 years [38–47]. However, fibre lasers provide pulse durations of about ~100 fs, which is much broader than that of crystal-based lasers. With special spectral broadening and pulse compression techniques, the pulse duration from fibre lasers keeps decreasing; and nowadays, a sub-10 fs level can be attained from an Er-doped fibre femtosecond laser. Table 5.4 shows an example laser specification for TPP.

Table 5.4. Summary of specifications for femtosecond laser for TPP-based AM.

Femtosecond Laser for TPP	
FemtoFibre pro-NIR	
✓ Wavelength	780 nm (fundamental wavelength: 1560 nm)
✓ Output power	140 mW (350 mW @ 1560 nm)
✓ Pulse width	100 fs (100 fs @ 1560 nm)
✓ Repetition rate	80 MHz (standard)
✓ Beam shape	TEM_{00}, $M^2 < 1.2$
✓ Beam divergence	1 mrad (2 mrad @ 1560 nm)
✓ Linear polarisation	95 % (horizontal)

TPP machine: Nanoscribe's Photonic Professional GT

The Photonic Professional GT provides a micro-fabrication platform for diverse scientific, technological and industrial applications. It enables the high-speed fabrication of micro-sized parts by providing a simple 3D printing workflow to convert designed 3D models into physical parts in various application fields; such as optics, medicine, fluidics or mechanics. As shown in Fig. 5.9, the Photonic Professional GT comprises four hardware sub-units, a NIR fibre femtosecond laser, beam-scanning galvano mirrors, a sample positioning stage and a monitoring microscope system. First, the laser unit is based on an Er-doped mode-locked fibre femtosecond laser (at 1550 nm) with an attached second harmonic generation part (to 775 nm). The average output power at 775 nm is over 140 mW and the pulse duration is ~100 fs. It provides linear polarisation with a good beam mode shape with the M^2 value of less than 1.2. Second, the galvano mirrors are similar in principle to the scanning mirrors in SLA. They enable high-speed beam scanning over 2D area but with lower positioning accuracy. Third, a sample positioning stage with piezoelectric transducers can provide nanometre resolution and accuracy. Its scanning speed is limited by the mass of the target sample, which is not the case for galvano mirrors. Therefore, for a high-speed printing over a relatively heavy sample, galvano mirrors are a better solution, whereas, for the high-precision printing of optical structures where the sample dimensions are critical, e.g. photonic crystals, diffraction gratings and plasmonic structures, piezo stages fit well. Fourth, a high-resolution microscopy system is helpful for real-time monitoring of the printing process. Depending on printing materials, the monitoring is not an easy task; it is because the colour, the intensity or the refractive index of the solidified photopolymer may not differ much from that of the liquid photopolymer. Special techniques, e.g. differential interference contrast (DIC) microscopy or phase contrast microscopy (PCM), could be applied in addition to the process monitoring. Table 5.5 shows the summary of Nanoscribe's Photonic Professional GT's specifications.

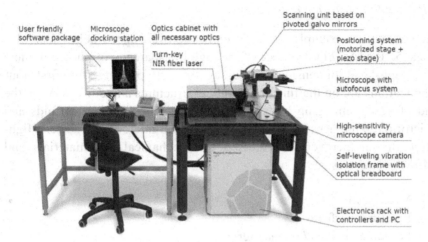

Fig. 5.9. The system configuration of TPP. (Courtesy of Nanoscribe GmbH)

Table 5.5. Summary of specifications of the Photonic Professional GT.

Nanoscribe's Photonic Professional GT Specifications		
Printing specifications	Galvano mode	Piezo mode
✓ Objective lens : 63×, NA=1.4		
✓ 3D lateral feature size (IP-DIP)	< 200 nm	< 200 nm
✓ 2D lateral resolution (IP-DIP)	< 500 nm	< 500 nm
✓ Range	200 μm (140×140 μm²)	300×300×300 μm³
✓ Writing speed	10 mm/s	100 μm/s
✓ Accessible writing area	up to 100×100 mm²	up to 100×100 mm²
Laser	NIR femtosecond laser	
Safety ✓ Electrical safety ✓ Laser safety	in accordance with EN 61010-1:2010 Class 1 for the whole system according to EN 60825-1:2007 (Class 3B for internal laser)	

TPP printed parts and applications

Representative examples of TTP applications are listed below (see also Figs. 5.10 and 5.11). Because of the high transparency of photopolymers and high-level dimensional accuracy of TPP printing, the first four applications are on the functional optical structures and devices. With the aid of newly emerging bio-compatible polymers, cellular scaffolds and microfluidic devices can be printed in sub-micrometre scale. High-strength polymers can be utilised for mechanical metamaterials and MEMS structures.

- Micro-optics
- Freeform surfaces
- Photonic crystals / metamaterials
- Photonic wire bonds
- Biomimetic / life sciences
- Tissue engineering / cell biology
- Micro- and nanofluidics
- Mechanical metamaterials
- MEMS

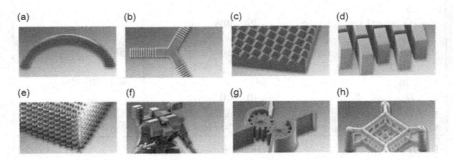

Fig. 5.10. TPP-based 3D printed parts. (a) Optical interconnect, (b) Photonic surfaces, (c) micro-optics, (d) Microfluidics, (e) Photonics, (f) Micro prototyping, (g) Mechanical microstructures, (h) Cell scaffolds and biomimetics. (Courtesy of Nanoscribe GmbH)

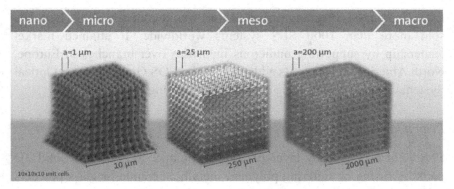

Fig. 5.11. From the micro- to the mesoscale: These lattice cubes were all printed using TPP to demonstrate the versatility of the micro production process for different scales. (Courtesy of Nanoscribe GmbH)

5.3. Powder-Based 3D Printing

This section describes the special group of solid-based 3D printing, which primarily utilises powder as the basic medium. Some of the systems in this group, such as Selective Laser Sintering (SLS) and Selective Laser Melting (SLM), are basically similar to the liquid-based SLA described in Chap. 2.1. SLS and SLM generally print the parts layer-by-layer like SLA, but the medium for building the 3D model are powders instead of photo-curable resin. The common feature of the systems described in this section is that the material used for 3D printing is powder-based.

5.3.1. *Selective Laser Sintering (SLS)*

History

SLS was first developed and patented by Dr Carl Deckard and academic adviser, Dr Joe Beaman at the University of Texas at Austin in the mid-1980s. Deckard and Beaman were involved in the start-up company DTM that worked on the design and development of SLS machines. In 2001, 3D Systems acquired DTM. The most recent patent on Deckard's SLS was issued on the 28 January 1997 and it expired on 28 January 2014.

EOS, another SLS leading manufacturer, was founded in 1989. EOS has sold more than 1000 SLS systems worldwide. It attained market leadership by supporting numerous industries over branches in Europe, North America and Asia. Its address is at the EOS GmbH Electro Optical Systems, Robert-String-Ring 1, D-82152 Krailling, Germany.

Principle: selective laser sintering

SLS is also a layer-by-layer printing technique. SLS creates solid 3D parts by fusing or sintering polymer, metallic, ceramic or glass powders using the heat energy generated by focusing a high-power laser beam onto the parts (see Fig. 5.12) [48,49]. Small powders consisting particles of tens of a micrometre in diameter are repeatedly spread over the existing previous layer. After the layer deposition, a high power laser beam scans over the surface and selectively sinters the powder together, which results in the cross-section of the printed product. After each cross-section is sintered, the powder bed is lowered by the layer thickness, a new layer is elevated to the top, and the beam scanning is repeated until the part is complete. SLS is based on a thermal sintering process using a laser beam as the energy source and so the photon energy from the laser should be efficiently converted into heat energy without a significant loss. Thus, material absorption is a critical issue in SLS; the selection of an appropriate laser system for the target material is the base step in SLS. The physical process can be full melting, partial melting or liquid-phase sintering. Since the finished part density in SLS depends on the peak power of the laser rather than its average power, SLS prefers pulsed lasers. Depending on the material, almost-perfect bulk density and material properties can be achieved. For higher throughputs, SLS preheats the powder bed below the Glass Transition Temperature (GTT). SLS does not require supporting structures or sacrificial layers, which is a key advantage compared to SLA. This is because, the residual un-sintered powders work to support the 3D printed structure. Thanks to this, highly complex geometries can simply be designed and manufactured without supporting structure concerns. If the target material can be made into powder form and can be sintered under laser focusing, SLS can work. Thus, SLS can produce parts from a wide range of commercially available materials. For example, SLS-printed polymer parts can be stiff, strong, water-tight, air-tight and/or heat-resistant by selecting the proper powder polymers. Two-component powder, either a coated powder or a powder mixture, is widely used in SLS. In the case of using single-component powders, the laser beam melts only the outer surface of the powders and sinters the solid non-melted cores to each other and also to the previous layer. SLS powders are generally produced by the ball milling process.

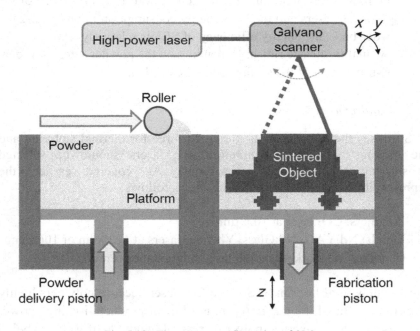

Fig. 5.12. The system configuration of SLS.

The strengths and weaknesses of SLS are listed as below.

Strengths
- Good part stability: Production-tough end-use parts and functional prototypes can be printed with a high resolution, high throughput and a low part cost.
- A wide range of processing materials: Any material in powder form can be sintered including nylon, polycarbonates, metals and ceramics.

- No part supports required: SLS allows users to take full advantage of 3D-printing's design freedom.
- Little post-processing required: Because of the absence of supporting structures, simply removing residual powders reveals the net shape.
- No post-printing curing required.

Weaknesses
- Large physical size of the unit.
- High power consumption: Due to the low wall-plug efficiency of the high-power lasers and the high energy consumption in preheating the powder-beds, the energy efficiency of SLS is relatively low.
- Low-quality surface finish: Depending on the powder size, additional post-processing of the surface could be needed.

Laser requirements for SLS system

SLS utilises the laser beam as the heat source for thermal sintering and the melting of powders; therefore, the lasers should be selected considering the materials' absorption. As covered earlier, the representative laser solutions for SLS are as follows.

- Polymers: CO_2 laser at 10.6 µm
- Metals: Nd:YAG, Nd:Glass, Yb-fibre lasers at 1030 nm or 1060 nm
- Ceramics: no market-leading laser solutions for ceramics.

Most widely used laser in SLS is the CO_2 laser because polymers highly absorb the infrared light at 10.6 µm. For higher productivity, high power CO_2 lasers with an average power of several hundreds of watt are being used nowadays.

SLS machine example: 3D Systems' SLS 500

3D Systems has manufactured several generations of the SLS system [49–56]. The current generation consists of the ProX 500 and sPro series printers. The ProX 500 is the largest SLS printer developed by 3D Systems (see Figs. 5.13 and 5.14). It uses DuraForm® ProX™ materials to produce high-quality prototypes and parts in various areas. The ProX

500 Plus prints additional materials, including glass filled and fibre-filled aluminium, and also offers faster and higher resolution modes. The included state-of-the-art Material Quality Centre (MQC) module ensures material recyclability for an efficient, clean and automated production. The sPro™ SLS system can be used as a flexible platform for printing different models to achieve different qualities and print speeds [57,58]. Table 5.6 summarises the specification of the ProX 500.

Fig. 5.13. 3D Systems' SLS 3D printing machine, ProX SLS 500. (Courtesy of 3D Systems)

Table 5.6. Summary of specifications of the ProX SLS 500.

3D Systems' ProX SLS 500 Specifications	
Standard Print Mode	
✓ Scanning speed	Fill: 12 m/s
	Outline: 3.5 m/s
✓ Layer thickness	Range: 0.08–0.15 mm
	Typical: 0.10 mm
✓ Volume build rate	2 litres/hour
High-resolution Print Mode	
✓ Scanning speed	N/A
✓ Layer thickness	N/A
✓ Volume build rate	N/A
High-speed Print Mode	
✓ Scanning speed	N/A
✓ Layer thickness	N/A
✓ Volume build rate	N/A
Print Envelope Capacity	381×330×457 mm³
Imaging System	ProScan DX Digital High Speed
Laser System	100 W / CO₂ laser

Fig. 5.14. Plastic parts printed by 3D Systems' ProX SLS 500. (Courtesy of 3D Systems)

Example of a SLS machine: 3D Systems' ProX DMP 320

3D Systems' DMP (Direct Metal Printing) series' process builds up dense and chemically-pure metal parts, by melting fine metallic powders with a laser beam. Layer thickness ranges between 5–30 microns, which results in a high resolution, good part accuracy and smooth surface finish.

The DMP adopts a high-power Yb-fibre laser over hundreds of watts as the energy source. ProX 400 can print the largest metal parts and tools, at over 500 mm × 500 mm × 500 mm in volume (see Fig. 5.15). The ProX 400 produces precision parts that are ready for assembly within a shorter time than dies and tooling can achieve (see Fig. 5.16 and Table 5.7) [59,60].

Fig. 5.15. 3D Systems' SLS 3D printing machine, ProX DMP 320. (Courtesy of 3D Systems)

Table 5.7. Summary of specifications of the ProX DMP 320.

3D Systems' ProX DMP 400 Specifications	
Laser Type	2 fibre lasers / 2 × 500 W
Laser Wavelength	1070 nm
Layer Thickness ✓ Range	Adjustable, min 10 μm, max 100 μm
Build Envelope Capacity	500×500×500 mm³
Ready-to-run Materials with Developed Print Parameters	Stainless steel, tool steel, non-ferrous alloys, super alloys and others
Minimum Feature Size	x=100 μm, y=100 μm, z=20 μm

When complex metal parts are needed fast, SLS of metal powders will show a competitive advantage in specialized industries like:

• Aerospace and defence
• Engine/component manufacturing
• Medical technology
• Patient-specific implants
• Dental applications
• Conformal cooling in tooling inserts

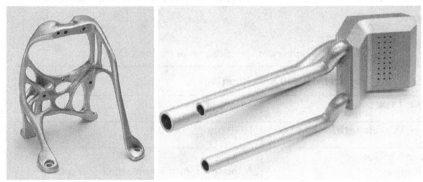

Fig. 5.16. Parts printed by 3D Systems' SLS/DMP. (Courtesy of 3D Systems)

SLS machines: EOS's EOSINT P760 and EOS M400 systems

EOSINT P760 is the world's first laser sintering system for 3D printing high-performance plastic products at high process temperatures (up to 385 °C) (see Fig. 5.17 and Table 5.8) [61]. For printing plastics with a high efficiency, two 50 W CO_2 lasers are installed as the energy source. During production, the integrated 'Online Laser Power Control (OLPC)' module monitors the laser's status for optimum and reproducible results on the printed components. For plastics, there are SLS product lines of FORMIGA P110, EOS P396, EOSINT P760 and EOSINT P800 in EOS [62].

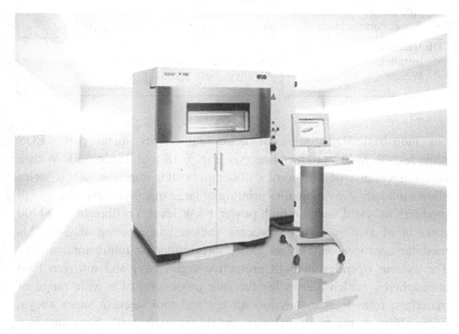

Fig. 5.17. EOS's SLS 3D printing machine, EOSINT P760. (Courtesy of EOS)

Table 5.8. Summary of specifications of the EOSINT P760.

EOS' EOSINT P760 Specifications	
Technical Data	
✓ Effective building volume	700 mm × 380 mm × 560 mm
✓ Building speed (material-dependent)	32 mm/h
✓ Layer thickness (material-dependent)	Typically 0.12 mm
✓ Support structure	Not required
✓ Laser type	CO_2: 2 EA × 50 W
✓ Precision optics	F-theta-lenses
✓ Scan speed during build process	Up to 2 × 6 m/s
✓ Power consumption	Maximum 12 kW / typically 3.1 kW
Dimensions	
✓ Installation space	4.8 m × 4.8 m × 3.0 m
✓ Weight	Approximately 2,300 kg

For industrial 3D printing of large high-quality metal parts, EOS provides the M400 product line (see Fig. 5.18 and Table 5.9). With a building volume of 400 mm × 400 mm × 400 mm, the M400 series allows industry-proven quality printing of large metallic parts with a high productivity yield, using a high power 1 kW level Yb-fibre laser. M400 consists of two stations, a process station and a setup station; this modular approach enables an easier integration for future innovations. The system operates in both protective argon (Ar) and nitrogen (N_2) atmospheres, which allows for the safe processing of a wide range of materials; from light metals to stainless and tool steel, to super alloys. EOS offers a range of powder metal materials with corresponding parameter sets that are optimised for the applications. For metals, there are SLS product lines of EOSINT M280, EOS M290, EOS M400 and PRECIOUS M080 in EOS.

Fig. 5.18. EOS's SLS 3D printing machine, EOS M400. (Courtesy of EOS)

Table 5.9. Summary of specifications of the EOS M400.

EOS M400 Specifications	
Technical Data	
✓ Effective building volume	400 mm × 400 mm × 400 mm
✓ Laser type	Yb-fibre laser, 1 kW
✓ Precision optics	F-theta-lenses
✓ Scan speed during build process	Up to 7.0 m/s
✓ Focus diameter	Approximately 90 μm
✓ Power consumption	Maximum 20.2 kW / typically 16.2 kW
✓ Building speed (material-dependent)	7 mm/h
Dimensions	
✓ System space	4.2 m × 1.6 m × 2.4 m
✓ Weight	Approximately 4,635 kg

5.3.2. *Selective Laser Melting (SLM)*

History

SLM started in 1995 at the Fraunhofer Institute ILT in Aachen, Germany. From the pioneering phase, Dr Dieter Schwarze and Dr Matthias Fockele from F&S Stereolithographietechnik GmbH collaborated with ILT researchers Dr Wilhelm Meiners and Dr Konrad Wissenbach. In the early 2000s, F&S started a partnership with MCP HEK GmbH which later on first changed its name to MTT Technology GmbH, before amending it to SLM Solutions GmbH [63].

Principle

Selective Laser Melting (SLM) is a 3D printing process that utilises a high-power laser beam to create 3D metal parts by fusing fine metal powders together [64–78]. SLM uses a comparable concept to the SLS in Chap. 5.3, but in SLM, the material is fully melted rather than partially sintered. This full melting allows different material properties, e.g. in the crystal structure, porosity and so on, which are different from the material properties offered by SLS. In SLM, a high-power laser beam selectively melts the thin layers of small metal powders evenly distributed over a substrate metal plate. Printing takes place inside a tightly-controlled gas chamber containing an inert gas (Ar or N_2) at oxygen levels below 500×10^{-6}. The laser beam is directed along the two lateral directions with two angle-scanning galvano mirrors. This beam scanning process is repeated layer after layer until the part is complete. Electron Beam Melting (EBM) is based on a similar printing process but it uses an electron beam as the energy source [79,80]. SLM manufacturing applications are being implemented in aerospace or medical orthopaedics.

The strengths and weaknesses of SLM are listed as below.

Strengths
- High-quality metal parts: Highly dense and bulk-metal-like printed 3D parts are guaranteed thanks to the full melting.

- Large range of metal materials: Base material is a single component powder (simpler in structure than SLS) so a wider range of materials can be used.
- Complex geometries: SLM produces tools and inserts with internal undercuts and channels for conformal cooling.
- Fast and low cost.
- High accuracy.

Weaknesses
- Large physical size of the unit.
- High power consumption: Due to low wall-plug efficiency of high-power lasers and high energy consumption in preheating the powder beds, the energy efficiency of SLM is relatively low. It is lower than SLS.
- Relatively slow process: Full melting of metals takes a longer time than metal sintering.
- Supports are required due to the heavy metallic powders.

Laser requirements for SLM system

SLM also utilises the laser beam as the heat source for the thermal melting process; therefore, the light absorption spectrum of metals should be considered when selecting the lasers. As covered earlier, the representative laser solutions for SLM are Nd:YAG and Yb-fibre lasers at 1030 nm or 1060 nm. Although most metals absorb at visible and ultraviolet wavelengths (compared to near-infrared around 1030 nm), there are no available high-power lasers at the visible and ultraviolet range to date. Nowadays, kilowatt-level high-power Yb-fibre lasers are used owing to their higher productivity (compared to other lasers).

SLM machine example: SLM Solutions' SLM 500 HL

SLM 500HL is a high-performance SLM platform providing a large build envelope and multi-beam technology (see Fig. 5.19 and Table 5.10). With the twin-beam (2×400 W) and quad-beam (4×400 W) configuration, this system is specifically designed for a high-throughput production environment [81]. A comprehensive monitoring and quality assurance system in SLM 500HL enables a high degree of process control. The types of materials which can be processed include stainless steel, tool steel, cobalt chrome, titanium and aluminium. All materials must exist in an atomised powder form and exhibit certain flow characteristics in order to be process-capable.

Fig. 5.19. SLM Solutions' SLM 3D printing machine, SLM 500 HL. (Courtesy of SLM Solutions)

Table 5.10. Summary of specifications of the SLM 500 HL.

SLM Solutions' SLM 500 HL Specifications	
Technical Data ✓ Build envelope ✓ 3D optics configuration ✓ Build rate ✓ Variable layer thickness ✓ Minimum feature size ✓ Beam focus diameter ✓ Maximum scan speed	700 mm × 380 mm × 560 mm 7 mm/h Typically 0.12 mm Not required
Dimensions ✓ Space ✓ Weight	5.2 m × 2.8 m × 2.7 m Approximately 3,100 kg (with powder) / 2400 kg (without powder)

SLM is well-suited for complex geometries and structures with thin walls and hidden voids/channels and with a low lot size. Early pioneering works with SLM were on lightweight parts for the aerospace industry where traditional manufacturing constraints, such as tooling and physical access to surfaces for machining, have restricted the design of components. In biomedical applications, SLM enables manufacturing hybrid forms where solid and lattice geometries can be produced together to create a single object, such as a hip stem or other orthopaedic implants where the integration is enhanced by the implant's surface geometry. Because SLM allows parts to be built additively to form near net-shape components, the material waste can be minimised.

SLM is being used for the manufacturing of dental models based on metal powders (see Fig. 5.20). It improved the productivity, quality and cost of dental products. Within a few hours, SLM can generate up to 400 tooth caps. Part accuracy, perfect fit, look and feel can be reproduced precisely.

SLM also enables a less expensive production of new tyre tread mould segments in a shorter time (see Fig. 5.20). SLM also allows tyre makers to create novel innovative designs for the next generation of tyres. Slits on a tyre tread can be further optimised with SLM.

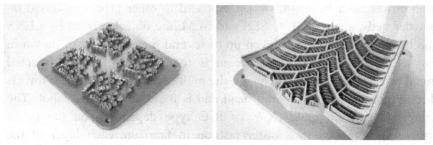

Fig. 5.20. SLM Solutions' SLM-printed parts. (Courtesy of SLM Solutions)

5.3.3. *Laser Engineering Net Shaping (LENS)*

History

Since 1997, Optomec Systems has produced hundreds of innovative 3D printer designs and products in the electronics, energy, life sciences, aerospace and defence markets, to name a few. With a strong track record of success and large-scale investments in product development,

Optomec has worked with leading companies and research organisations in the field of 3D printing. The patented 'Laser Engineered Net Shaping' (LENS) and 'Aerosol Jet' technologies deliver the benefits of 3D printing with new levels of efficiency and performances. This book focuses on LENS because Aerosol Jet is not based on a laser. The first commercial LENS 3D printer was delivered in 1998—and it won the "Top 25 Technologies of Year" award that same year. In 1999, Optomec was awarded a $9.0 million DARPA contract and another $5.0 million LENS project with Boeing, Rolls-Royce, Siemens and the US Air Force, Army and Navy in 2003. The 3rd generation LENS 3D printer was released in 2003. LENS has been used especially in the development and repair of military/aerospace parts. Newly emerging applications are in the bio sciences, health sciences and energy.

Principle

LENS is a 3D printing technology developed for fabricating metal parts by using a metal powder injected into a melt pool created by a focused, high-power laser beam [82–84]. Compared to other processes based on powder beds, such as SLA, SLS and SLM, the object printed by LENS can be substantially larger, even up to several metres long. As shown in Fig. 5.21, a high power laser beam is focused onto the side-injected metal powder to melt and generate the melt pool. The laser beam travels through the centre of the print head and is focused to a small spot. The sample translation table (X-Y or R-Θ type depending on the target structure) is moved in a raster fashion to fabricate each layer of the object. The head is then shifted up vertically after each layer is printed. Metal powders can be delivered using the gravitational force or injected by the aid of pressurised carrier gas. An inert gas is usually used to shield the melt pool from oxygen (an oxygen level of less than 10 ppm is required) for better process control and to promote layer to layer adhesion by providing better surface wetting. Due to the small melt pool and high stage travel speed, the deposited part cools very fast (up to 10,000 °C/s), which results in the generation of fine grain structures. The grain size is about one order of magnitude smaller than the wrought bulk products'. Therefore, the printed parts' mechanical properties could be better than castings and bulk metals. LENS 3D printers are well-suited to fabricate, enhance and repair high-performance metal components used in aerospace, defence, power generation, and medical device manufacturing industries.

LENS is similar to SLM, but the metal powder is only applied where the material is being added to the part at that moment, without the need for the powder bed. LENS can print a wide range of metals, including titanium, stainless steel, aluminium, other specialty materials, such as composites and functionally graded materials.

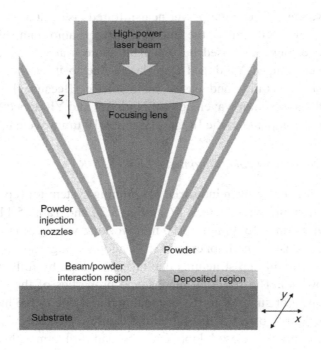

Fig. 5.21. The system configuration of LENS.

The strengths and weaknesses of LENS are listed as below.

Strengths
- Superior material properties: Comparable or even better properties.
- Reduced post-processing requirements
- Complex parts

Weaknesses
- Limited materials: Only metal parts
- Large physical unit size: Relatively large area is required for the unit
- High power consumption

Laser requirements for LENS system

LENS system utilises high-power near-infrared lasers at a wavelength of 1030 nm or 1060 nm as the energy source. Lamp-pumped Nd:YAG lasers have long been used but the light sources are being changed to diode-laser-pumped Yb-doped fibre lasers due to its higher efficiency, high system stability and easier maintenance. Because LENS is for metals, the absorption wavelength of Nd:YAG and Yb-doped fibre lasers match well compared to the IR CO_2 laser at a 10 μm wavelength.

LENS machine example: Optomec's LENS 850-R

LENS 850-R is a proven industrial 3D printing system for repair, rework, modification and manufacturing (see Fig. 5.22 and Table 5.11). It offers a large 900 mm × 1500 mm × 900 mm working volume, making it ideal for large industrial components [85]. It adopts a high-power Yb-doped fibre laser to build up structures one layer at a time by fully melting the metal powders. The resulting mechanical properties of the material are equivalent to or superior to the original material due to the high cooling rate. The 850-R offers a full range of features, including 5-axis CNC-controlled motion, closed loop controls and full atmosphere control. Optomec's 3D Printers are designed to work in harmony within existing subtractive or formative production processes, allowing manufacturers to 3D-print within their factory.

Fig. 5.22. Optomec's LENS 3D printing machine, LENS 850-R. (Courtesy of Optomec)

Table 5.11. Summary of specifications of the LENS 850-R.

Optomec's LENS 850-R Specifications	
Technical Data	
✓ Process work volume	900 mm × 1500 mm × 900 mm
✓ Enclosure	Class I laser enclosure, Hermetically sealed to maintain process environment and Safety
✓ Motion control	5-axes standard: XYZ linear gantry motion Tilt-Rotate worktable All axes under full CNC control
✓ Position accuracy	± 0.25 mm
✓ Linear resolution	± 0.025 mm
✓ Motion velocity	60 mm/s
✓ Deposition rate	Up to 0.5 kg/hr
✓ Parts handling	Tilt-Rotate table tilts ±90°, infinite rotation. Rails and part cart allow the table to move through the machine and out. 380 mm diameter antechamber.
✓ Gas purification system	Dual unit maintains 02 level continuously < 10 ppm
✓ Powder feeder	Two feeders each hold up to 14 kg of powder
✓ Lasers	1 or 2 kW IPG fibre laser
✓ Enclosure dimensions	3 m × 3 m × 3 m w/o gas purification system or laser

LENS is known to provide a special solution for product repair. The repair of high-value metal components is important for increasing product lifespans, reducing operating costs and maintaining a high level of readiness. Compared to the traditional repair processes, LENS is a highly selective method that precisely adds material to the damaged areas with minimal heat effect (see Fig. 5.23). Therefore, LENS has enabled the repair of the thin-walled components such as the blade and other structures in gas turbine engines. The resulting LENS repairs have mechanical properties that can be equivalent or even superior to wrought materials. The ability of the LENS system to add material to existing components makes it an essential system for service and repair applications.

Fig. 5.23. Optomec's LENS application example in turbine blade repair. (a) T700 stage 1 blisk before LENS leading edge repair process (b) The blisk after finishing the process. (Courtesy of Optomec)

5.4. Solid-Based 3D Printing

Solid-based 3D printing systems are very different from the liquid-based or powder-based systems. They utilise solids (in one form or another) as the primary medium to create the 3D model. Because the solid base material needs to be cut or sectioned, they can be regarded to be located at the borderline between additive and subtractive processes.

5.4.1. *Laminated object manufacturing (LOM)*

History

LOM is a 3D printing technique commercialised by Helisys Inc. LOM pioneer Helisys sold more than 375 systems during the 1990s, yet folded their business in 2000. It has been succeeded by Cubic Technologies, which provides parts for and service old Helisys machines. Using a CO_2 laser, Helisys's LOM printer cuts successive cross sections out of adhesive-coated paper. They struggled with reliability; an issue from which they could not fully recover. Other companies specialising in LOM have also closed their doors in recent years, including the Israel-based Solido 3D and Kira, Inc. in Japan.

Mcor Technologies Ltd. brought 'Selective Deposition Lamination (SDL)', which has some similarities with LOM, into the market. SDL can be distinguished from LOM by its inter-layer adhesion method. In SDL, the printer deposits adhesive drops only on the area which would become the part; enabling the fast and easy excavation of the part from its supports when printing is complete. In the previous LOM, the adhesive was pre-applied over the entire material surface; therefore, everything was glued together after the process, including the supports around the model. This made the excavation of the printed model difficult and often resulted in a part breakage. Mcor Technologies' SDL cuts each layer of paper using a tungsten carbide blade instead of the laser, so it is out of this book's scope.

Principle

In LOM, layers of adhesive-coated solid material, such as paper, plastic or metal laminates, are successively glued together and cut into the shape with a laser cutter or a knife [86–88]. Traditional LOM followed this procedure (see also Fig. 5.24). First, the sheet is adhered to a substrate by a heated roller. Second, a laser beam hatches the non-part area for subsequent waste removal. Third, the platform is elevated down. Fourth, a fresh sheet layer is rolled out into position. Between laser and knife cutting, the laser cutting is basically a non-contact and point-based cutting process, and so has a comparative advantage in printing high-resolution complex geometries. Lateral resolution is fundamentally limited by the optical diffraction limit to tens of micrometres, and layer thickness resolution is determined by the material feed which ranges in thickness from one to a few sheets of copy paper. LOM's dimensional accuracy can be slightly better than SLA, SLS or SLM. Relatively large volumes can be printed thanks to the simplicity of the process.

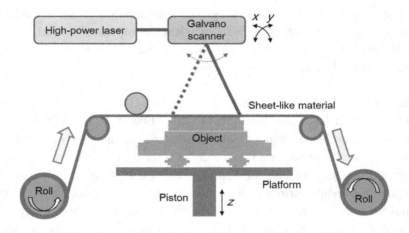

Fig. 5.24. A LOM system's configuration.

Strengths and weaknesses of LOM are listed as below.

Strengths
- Wide variety of materials: Any material in sheet form can be used
- Fast build time: No melting or chemical process involved.
- High precision: Lateral precision is determined by the optical diffraction limit. In the depth direction, the precision is defined by each layer's thickness.
- Support structure: Residual material part works as the support.
- Post-curing: No physical or chemical changes.

Weaknesses

- Precise power adjustment: Previous layers can be cut through
- Fabrication of thin walls: Relatively low rigidity in z-direction
- Integrity of prototypes: Low strength due to limited adhesives
- Removal of supports

Laser requirements for LOM system

LOM does not require high-performance lasers. If the focused laser beam can cut the sheet material into pieces, that is good enough. Therefore, cost is the most important criteria in laser selection in LOM. Consequently, less expensive CO_2 lasers have been widely used, with printing papers and/or polymers as the main sheet material. Other types of lasers can also be applied to LOM.

LOM machine example: Helisys' LOM

Currently, there is no commercially-available laser-based LOM 3D printer on the market. Table 8.12 shows the discontinued Helisys' LOM system's specifications.

Table 5.12. Summary of specifications of Helisys' LOM.

Helisys' LOM Specifications	
Technical Data	
✓ Maximum part size	812 mm × 559 mm × 508 mm
✓ Minimum feature size	200 μm
✓ Minimum layer thickness	50 μm
✓ Tolerance	100 μm
✓ Surface finish	Rough
✓ Build speed	Fast
✓ Materials	Thermoplastics such as PVC, paper, composites

Problems

1. List five examples of different terms for 3D printing.

2. List laser-based manufacturing processes that can be categorised as:

   a. additive manufacturing (Four)

   b. subtractive manufacturing (Three) and

   c. formative manufacturing (Three).

3. Explain the printing steps of stereolithography (SLA). What kind of laser is used in SLA? What is the reason? What is the average power of the laser?

4. What is the function of the galvano scanner in SLA?

5. List five strengths and three weaknesses of SLA.

6. Explain the function of the digital micro-mirror device (DMD) in DLP. What is the number of micromirrors in a DMD? Explain the DLP printing resolution based on DMD's pixel number.

7. What are the printing volume and the voxel resolution in DLP?

8. List five strengths and three weaknesses of DLP.

9. Explain two representative applications of DLP.

10. Explain what two-photon polymerisation (TPP) is. Compare why TPP is different from photo-polymerisation in SLA. What is the advantage of using TPP compared to photo-polymerisation?

11. What are the achievable voxel resolution and printing volume in TPP?

12. What kind of laser is used in TPP? What is the reason? What are the pulse duration, average power and repetition rate of the laser?

13. Why are fibre lasers getting wide-spread in 3D printing?

14. List three strengths and three weaknesses of TPP.

15. Explain the principle of selective laser sintering (SLS) compared to SLA. What are the similarities and differences in the two techniques?

16. What are the achievable voxel resolution and printing volume in SLS? What fundamentally determines these?

17. Which kind of laser is used in SLS? Depending on the material, which parameters of the lasers need to be changed?

18. List five strengths and three weaknesses of SLS.

19. Explain the principle of selective laser melting (SLM) compared to SLA and SLS. What are the similarities and differences?

20. What are the achievable voxel resolution and printing volume in SLM? What fundamentally determines these?

21. Which kind of laser is used in SLM? What is the required average power of the laser?

22. List five strengths and three weaknesses of SLM.

23. Explain two representative applications of SLM.

24. Explain the principle of laser engineered net shaping (LENS) compared to SLM. What are are the similaries and differences?

25. List three strengths and three weaknesses of LENS.

26. Explain two representative applications of LENS.

References

[1] Chua, C. K., & Leong, K. F. (2014). *3D printing and additive manufacturing: principles and applications: principles and applications.* World Scientific, Singapore.

[2] Chua, C. K., & Leong, K. F. (2010). *Rapid Prototyping: Principles and Applications.* 3rd edition. World Scientific, Singapore.

[3] Chua, C. K., Chou, S. M., & Wong, T. S. (1998). *A study of the state-of-the-art rapid prototyping technologies*. The International Journal of Advanced Manufacturing Technology, **14**(2), 146–152.

[4] Chua, C. K., Feng, C., Lee, C. W., & Ang, G. Q. (2005). *Rapid investment casting: direct and indirect approaches via model maker II*. The International Journal of Advanced Manufacturing Technology, **25**(1–2), 26–32.

[5] Lipson, H., & Kurman, M. (2013). *Fabricated: The new world of 3D printing*. John Wiley & Sons.

[6] Chua, C. K, & Yeong, W. Y. (2014). *Bioprinting: principles and applications*. World Scientific, Singapore.

[7] Tan, J. Y., Chua, C. K., & Leong, K. F. (2010). *Indirect fabrication of gelatin scaffolds using rapid prototyping technology*. Virtual and Physical Prototyping, **5**(1), 45–53.

[8] Gibson, I., Rosen, D., & Stucker, B. (2014). *Additive manufacturing technologies: 3D printing, rapid prototyping, and direct digital manufacturing*. Springer.

[9] Burns, M. (1994). Research Notes, *Rapid Prototyping Report*, **4**(3), 3–6.

[10] Burns, M. (1993). *Automated Fabrication*. New Jersey: PTR Prentice Hall.

[11] Jacobs, P. F. (1992). *Rapid prototyping & manufacturing: fundamentals of stereolithography*. Society of Manufacturing Engineers.

[12] Chua, C. K. (1994). *Three-dimensional rapid prototyping technologies and key development areas*. Computing & Control Engineering Journal, **5**(4), 200–206.

[13] Chua, C. K., Chou, S. M., & Wong, T. S. (1998). *A study of the state-of-the-art rapid prototyping technologies*. The International Journal of Advanced Manufacturing Technology, **14**(2), 146–152.

[14] Jacobs, P. F. (1995). *Stereolithography and other RP&M technologies: from rapid prototyping to rapid tooling*. Society of Manufacturing Engineers.

[15] Cooke, M. N., Fisher, J. P., Dean, D., Rimnac, C., & Mikos, A. G. (2003). *Use of stereolithography to manufacture critical-sized 3D biodegradable scaffolds for bone ingrowth*. Journal of Biomedical Materials Research Part B: Applied Biomaterials, **64**(2), 65–69.

[16] Hutmacher, D. W., Sittinger, M., & Risbud, M. V. (2004). Scaffold-based tissue engineering: rationale for computer-aided design and solid free-form fabrication systems. Trends in Biotechnology, **22**(7), 354–362.

[17] Griffith, M. L., & Halloran, J. W. (1996). *Freeform fabrication of ceramics via stereolithography*. Journal of the American Ceramic Society, **79**(10), 2601–2608.

[18] Melchels, F. P., Feijen, J., & Grijpma, D. W. (2010). *A review on stereolithography and its applications in biomedical engineering.* Biomaterials, **31**(24), 6121–6130.

[19] Wilson, J. E. (1974). *Radiation Chemistry of Monomers, Polymers, and Plastics.* Marcel Dekker, NY.

[20] Lawson, K. (1994). UV/EB Curing in North America, *Proceedings of the International UV/EB Processing Conference*, Florida, USA, May 1–5, 1.

[21] Reiser, A. (1989) *Photosensitive Polymers.* John Wiley, NY.

[22] 3D Systems Product brochure. (2016). SLA Production series.

[23] 3D Systems Product brochure. (2016). ProJet® 6000, 7000, 3500, 3510, 5000 and 5500X.

[24] Hendrick, J. Perfactory® - A Rapid Prototyping system on the way to the "personal factory" for the end user. Envision Technologies GmbH.

[25] Ho, C. M. B., Ng, S. H., Li, K. H. H., & Yoon, Y. J. (2015). *3D printed microfluidics for biological applications.* Lab on a Chip, **15**(18), 3627–3637.

[26] A. Lifton, V., Lifton, G., & Simon, S. (2014). *Options for additive rapid prototyping methods (3D printing) in MEMS technology.* Rapid Prototyping Journal, **20**(5), 403–412.

[27] Comina, G., Suska, A., & Filippini, D. (2014). *Low cost lab-on-a-chip prototyping with a consumer grade 3D printer.* Lab on a Chip, **14**(16), 2978–2982.

[28] O'Neill, P. F., Azouz, A. B., Vazquez, M., Liu, J., Marczak, S., Slouka, Z., ... & Brabazon, D. (2014). *Advances in three-dimensional rapid prototyping of microfluidic devices for biological applications.* Biomicrofluidics, **8**(5), 052112.

[29] Maruo, S., Nakamura, O., & Kawata, S. (1997). *Three-dimensional microfabrication with two-photon-absorbed photopolymerisation.* Optics Letters, **22**(2), 132–134.

[30] Cumpston, B. H., Ananthavel, S. P., Barlow, S., Dyer, D. L., Ehrlich, J. E., Erskine, L. L., ... & Qin, J. (1999). *Two-photon polymerisation initiators for three-dimensional optical data storage and microfabrication.* Nature, **398**(6722), 51–54.

[31] Kawata, S., Sun, H. B., Tanaka, T., & Takada, K. (2001). *Finer features for functional microdevices.* Nature, **412**(6848), 697–698.

[32] Gattass, R. R., & Mazur, E. (2008). *Femtosecond laser micromachining in transparent materials.* Nature Photonics, **2**(4), 219–225.

[33] Xiong, W., Zhou, Y. S., He, X. N., Gao, Y., Mahjouri-Samani, M., Jiang, L., ... & Lu, Y. F. (2012). Simultaneous additive and subtractive three-dimensional

nanofabrication using integrated two-photon polymerisation and multiphoton ablation. Light: Science & Applications, 1(4), e6.

[34] Obata, K., El-Tamer, A., Koch, L., Hinze, U., & Chichkov, B. N. (2013). *High-aspect 3D two-photon polymerisation structuring with widened objective working range (WOW-2PP)*. Light: Science & Applications, 2(12), e116.

[35] Fabian, N., & Hermatschweiler. M. (2014). *Additive Manufacturing of Micro-sized Parts*. Laser Technik Journal **11.5**, 16–18.

[36] Jin, J., Kim, Y. J., Kim, Y., Kim, S. W., & Kang, C. S. (2006). *Absolute length calibration of gauge blocks using optical comb of a femtosecond pulse laser*. Optics Express, **14**(13), 5968–5974.

[37] Kim, Y. J., Jin, J., Kim, Y., Hyun, S., & Kim, S. W. (2008). *A wide-range optical frequency generator based on the frequency comb of a femtosecond laser*. Optics Express, **16**(1), 258–264.

[38] Kim, Y., Kim, S., Kim, Y. J., Hussein, H., & Kim, S. W. (2009). Er-doped fiber frequency comb with mHz relative linewidth. Optics express, **17**(14), 11972–11977.

[39] Kim, Y., Kim, Y. J., Kim, S., & Kim, S. W. (2009). *Er-doped fiber comb with enhanced fceo S/N ratio using Tm: Ho-doped fiber*. Optics Express, **17**(21), 18606–18611.

[40] Kim, Y. J., Kim, Y., Chun, B. J., Hyun, S., & Kim, S. W. (2009). *All-fiber-based optical frequency generation from an Er-doped fiber femtosecond laser*. Optics Express, **17**(13), 10939–10945.

[41] Lee, J., Kim, Y. J., Lee, K., Lee, S., & Kim, S. W. (2010). *Time-of-flight measurement with femtosecond light pulses*. Nature Photonics, **4**(10), 716–720.

[42] Kim, S., Kim, Y., Park, J., Han, S., Park, S., Kim, Y. J., & Kim, S. W. (2012). *Hybrid mode-locked Er-doped fiber femtosecond oscillator with 156 mW output power*. Optics Express, **20**(14), 15054–15060.

[43] Jang, Y. S., Lee, J., Kim, S., Lee, K., Han, S., Kim, Y. J., & Kim, S. W. (2014). *Space radiation test of saturable absorber for femtosecond laser*. Optics Letters, **39**(10), 2831–2834.

[44] Kim, Y. J., Coddington, I., Swann, W. C., Newbury, N. R., Lee, J., Kim, S., & Kim, S. W. (2014). *Time-domain stabilization of carrier-envelope phase in femtosecond light pulses*. Optics Express, **22**(10), 11788–11796.

[45] Kim, S., Park, J., Han, S., Kim, Y. J., & Kim, S. W. (2014). *Coherent supercontinuum generation using Er-doped fiber laser of hybrid mode-locking*. Optics Letters, **39**(10), 2986–2989.

[46] Lee, J., Lee, K., Jang, Y. S., Jang, H., Han, S., Lee, S. H., ... & Kim, S. W. (2014). *Testing of a femtosecond pulse laser in outer space.* Scientific Reports, **4**, 5134.

[47] Jang, H., Jang, Y. S., Kim, S., Lee, K., Han, S., Kim, Y. J., & Kim, S. W. (2015). *Polarization maintaining linear cavity Er-doped fiber femtosecond laser.* Laser Physics Letters, **12**(10), 105102.

[48] Gibson, I., & Shi, D. (1997). *Material properties and fabrication parameters in selective laser sintering process.* Rapid Prototyping Journal, **3**(4), 129–136.

[49] Agarwala, M., Bourell, D., Beaman, J., Marcus, H., & Barlow, J. (1995). *Direct selective laser sintering of metals.* Rapid Prototyping Journal, **1**(1), 26–36.

[50] Liao, H. T., Chang, K. H., Jiang, Y., Chen, J. P., & Lee, M. Y. (2011). *Fabrication of tissue engineered PCL scaffold by selective laser-sintered machine for osteogeneisis of adipose-derived stem cells: the research has proven that a bone tissue-engineered scaffold can be made using the selective laser sintering method.* Virtual and Physical Prototyping, **6**(1), 57–60.

[51] Jhabvala, J., Boillat, E., Antignac, T., & Glardon, R. (2010). *On the effect of scanning strategies in the selective laser melting process.* Virtual and Physical Prototyping, **5**(2), 99–109.

[52] Ponche, R., Hascoët, J. Y., Kerbrat, O., & Mognol, P. (2012). *A new global approach to design for additive manufacturing: A method to obtain a design that meets specifications while optimizing a given additive manufacturing process is presented in this paper.* Virtual and Physical Prototyping, **7**(2), 93–105.

[53] Oxman, N., Tsai, E., & Firstenberg, M. (2012). Digital anisotropy: A variable elasticity rapid prototyping platform: This paper proposes and demonstrates a digital anisotropic fabrication approach by employing a multi-material printing platform to fabricate materials with controlled gradient properties. Virtual and Physical Prototyping, 7(4), 261–274.

[54] De Campos, B. M., Bandeira, L. C., Calefi, P. S., Ciuffi, K. J., Nassar, E. J., Silva, J. V. L., ... & Maia, I. A. (2011). *Protective coating materials on nylon substrate by sol–gel: The sol-gel process can be used to coat a nylon substrate, changing the properties of the material and the interaction between the coating and the substrate takes place through the NH groups of nylon.* Virtual and Physical Prototyping, **6**(1), 33–39.

[55] Nakthewan, K., & Koomsap, P. (2013). *Direct contour generation for structured light system-based selective data acquisition: This paper reports a direct method to extract contour information based on the surface data acquired via structured light system.* Virtual and Physical Prototyping, **8**(2), 135–163.

[56]   Salmoria, G. V., Klauss, P., Zepon, K., Kanis, L. A., Roesler, C. R. M., & Vieira, L. F. (2012). *Development of functionally-graded reservoir of PCL/PG by selective laser sintering for drug delivery devices: This paper presents a selective laser sintering-fabricated drug delivery system that contains graded progesterone content.* Virtual and Physical Prototyping, **7**(2), 107–115.

[57]   3D Systems Product Brochure. (2016). SLS® Systems.

[58]   3D Systems Product Brochures and Datasheets.

[59]   3D Systems. (2005). SLS® Technology Featured at the "Extreme Textiles" Exhibition at the Smithsonian's Cooper-Hewitt, National Design Museum.

[60]   3D Systems. (2014). Customer Success Story – Hankook Tire.

[61]   EOS GmbH. (2013). The Challenge of Customised Products.

[62]   EOS GmbH Product Brochures and Datasheets.

[63]   Sturm, R., Erickson-Harris, L., & Onge, D. S. (2002). SLM Solutions: A Buyer's Guide. Enterprise Management Associates.

[64]   Liu, A., Chua, C. K., & Leong, K. F. (2010). *Properties of test coupons fabricated by selective laser melting.* In Key Engineering Materials (Vol. 447, pp. 780–784). Trans Tech Publications.

[65]   Niendorf, T., Leuders, S., Riemer, A., Richard, H. A., Tröster, T., & Schwarze, D. (2013). *Highly anisotropic steel processed by selective laser melting.* Metallurgical and Materials Transactions B, **44**(4), 794–796.

[66]   Liu, Z. H., Chua, C. K., & Leong, K. F. (2012). Heat treatment of SLM M2 high speed steel parts. *In 5th International Conference PMI*, Ghent, Belgium.

[67]   Zhang, D. Q., Liu, Z. H., & Chua, C. K. (2013, September). Investigation on forming process of copper alloys via Selective Laser Melting. In High Value Manufacturing: Advanced Research in Virtual and Rapid Prototyping: *Proceedings of the 6th International Conference on Advanced Research in Virtual and Rapid Prototyping*, Leiria, Portugal, 1–5 October, 2013 (p. 285). CRC Press.

[68]   Leuders, S., Thöne, M., Riemer, A., Niendorf, T., Tröster, T., Richard, H. A., & Maier, H. J. (2013). *On the mechanical behaviour of titanium alloy TiAl6V4 manufactured by selective laser melting: Fatigue resistance and crack growth performance.* International Journal of Fatigue, **48**, 300–307.

[69]   Loh, L. E., Chua, C. K., Yeong, W. Y., Song, J., Mapar, M., Sing, S. L., ... & Zhang, D. Q. (2015). *Numerical investigation and an effective modelling on the Selective Laser Melting (SLM) process with aluminium alloy 6061.* International Journal of Heat and Mass Transfer, **80**, 288–300.

[70]   Jhabvala, J., Boillat, E., Antignac, T., & Glardon, R. (2010). *On the effect of scanning strategies in the selective laser melting process.* Virtual and Physical Prototyping, **5**(2), 99–109.

[71]   Averyanova, M., Bertrand, P., & Verquin, B. (2011). *Manufacture of Co-Cr dental crowns and bridges by selective laser Melting technology: This paper presents the successful application of the selective laser melting technology in dental frameworks manufacturing from Co-Cr alloy using Phenix PM 100T Dental Machine over a production period of 14 months.* Virtual and Physical Prototyping, **6**(3), 179–185.

[72]   Yasa, E., Deckers, J., Kruth, J. P., Rombouts, M., & Luyten, J. (2010). *Charpy impact testing of metallic selective laser melting parts.* Virtual and Physical Prototyping, **5**(2), 89–98.

[73]   Delgado, J., Ciurana, J., & Serenó, L. (2011). *Comparison of forming manufacturing processes and selective laser melting technology based on the mechanical properties of products: In this work, the superior property of the selective laser melting technology is presented by comparing four real parts manufactured using forming processes and selective laser melting technology and analysed for tension, compression and flexural.* Virtual and Physical Prototyping, **6**(3), 167–178.

[74]   Averyanova, M., Bertrand, P., & Verquin, B. (2011). *Studying the influence of initial powder characteristics on the properties of final parts manufactured by the selective laser melting technology: A detailed study on the influence of the initial properties of various martensitic stainless steel powders on the final microstructures and mechanical properties of parts manufactured using an optimized SLM process is reported in this paper.* Virtual and Physical Prototyping, **6**(4), 215–223.

[75]   Papadakis, L., Loizou, A., Risse, J., Bremen, S., & Schrage, J. (2014). *A computational reduction model for appraising structural effects in selective laser melting manufacturing: A methodical model reduction proposed for time-efficient finite element analysis of larger components in Selective Laser Melting.* Virtual and Physical Prototyping, **9**(1), 17–25.

[76]   Gu, D., & Zhang, G. (2013). *Selective laser melting of novel nanocomposites parts with enhanced tribological performance: nanocrystalline TiC/Ti nanocomposites parts were built via SLM technology and the densification, microstructures, microhardness and tribological performance were investigated.* Virtual and Physical Prototyping, **8**(1), 11–18.

[77] Yadroitsev, I., & Yadroitsava, I. (2015). *Evaluation of residual stress in stainless steel 316L and Ti6Al4V samples produced by selective laser melting.* Virtual and Physical Prototyping, **10**(2), 67–76.

[78] Wen, S. F., Yan, C. Z., Wei, Q. S., Zhang, L. C., Zhao, X., Zhu, W., & Shi, Y. S. (2014). *Investigation and development of large-scale equipment and high performance materials for powder bed laser fusion additive manufacturing: This paper reports a uniform preheating technique, a multi-laser scanning technique and a technique to prepare nylon coated composite powder.* Virtual and Physical Prototyping, **9**(4), 213–223.

[79] Arcam AB. (2007). Electron Beam Melting.

[80] Arcam AB. *New orthopedic implants improve people's quality of life.* Rapid News.

[81] SLM Solutions Product Brochures and Datasheets.

[82] Griffith, M. L., Keicher, D. M., Atwood, C. L., Romero, J. A., Smugeresky, J. E., Harwell, L. D., & Greene, D. L. (1996, August). Free form fabrication of metallic components using laser engineered net shaping (LENS). *In Proceedings of the Solid Freeform Fabrication Symposium* (pp. 125-131). Austin, TX: University of Texas at Austin.

[83] Atwood, C., Griffith, M. L., Schlienger, M. E., Harwell, L. D., Ensz, M. T., Keicher, D. M., ... & Smugeresky, J. E. (1998, November). Laser engineered net shaping (LENS): a tool for direct fabrication of metal parts. *In Proceedings of ICALEO* (Vol. 98, pp. 16-19).

[84] Liu, W., & DuPont, J. N. (2003). *Fabrication of functionally graded TiC/Ti composites by laser engineered net shaping.* Scripta Materialia, **48**(9), 1337–1342.

[85] Optomec Product Brochures and Datasheets.

[86] Mueller, B., & Kochan, D. (1999). *Laminated object manufacturing for rapid tooling and patternmaking in foundry industry.* Computers in Industry, **39**(1), 47–53.

[87] Zhang, Y., He, X., Du, S., & Zhang, J. (2001). *Al₂O₃ ceramics preparation by LOM (laminated object manufacturing).* The International Journal of Advanced Manufacturing Technology, **17**(7), 531–534.

[88] Park, J., Tari, M. J., & Hahn, H. T. (2000). *Characterization of the laminated object manufacturing (LOM) process.* Rapid Prototyping Journal, **6**(1), 36–50.

Chapter 6

ADVANCED 3D MANUFACTURING: MICRO & NANOSCALE PATTERNING

Patterning and printing 2D and 3D structures on different materials, which also include photosensitive materials, found tremendous interest among the scientific and industrial communities. Laser interference based concepts and methodology can make a significant impact in this area. In this context, the main aim of this chapter is to give the fundamental aspects of patterning at the micro and nanoscale dimensions using the laser interference principle. It is to be emphasised here that these patterning concepts can be employed to print 2D and 3D structures on different materials. The chapter begins with an introduction and covers the fundamentals of laser interference. It then details patterning based on 2–beam and 4-beam interference with the help of a multiple beam interference lithography system. The effects of intensity and polarisation of the interfering beam on the structures are also detailed in this chapter. And finally, it finishes by covering ultrafast laser machining.

6.1 Introduction

It is well known that optical interference concepts and methodologies could be successfully applied to fabricate periodic features in one and two dimensions. A one-dimensional periodic fringe pattern, which has a sinusoidal intensity distribution, could be formed by interfering two waves satisfying coherence requirements. This method can be extended to produce two-dimensional periodic patterns by exposing the same material to a two-beam interference pattern more than one time, at different angles. In general, a double exposure with a 90° sample rotation yields a 2D pattern with a square symmetry, whereas a triple exposure with a 120° sample rotation between exposures results in hexagonal lattice features [1,2]. Another method for the fabrication of 2D and 3D periodic patterns is to employ a large number of coherent beams for interference [3,4]. Four coherent plane waves could be used to fabricate periodic patterns in a square symmetry, where the polarisation of individual beams has a significant effect on the image contrast and period. Such multiple-beam interference techniques have been widely used in the fabrication of a variety of 2D and 3D periodic structures at the micro and nano-scale [5–8]. Various configurations involving multiple beams have been successfully employed to fabricate different types of three-dimensional periodic structures [9–11]. A minimum of four beams are required for 3D periodic pattern generation by interference. Out of the various 3D periodic structures, the one with a diamond like lattice is of particular interest as it is reported to have a full band gap of around 25% of the centre frequency [12–16]. Various other 3D structures are reported to use different interference pattern exposures [13, 17].

The main aim of this chapter is to give the fundamental aspects of patterning at the micro and nanoscale dimensions using the laser interference principle. It is to be emphasised here that these patterning concepts can be employed to print 3D structures layer by layer; which will allow the realising of different 3D-printed structures on different materials, which is an area of potential future research and development.

In this context, this chapter aims to provide the fundamentals of interference and patterning using laser interference (Laser Interference Lithography) and the effects of various parameters such as intensity, phase and polarisation.

6.2 Interference

When two or more light beams are allowed to cross each other, there may be a modification to the intensity (subject to certain conditions) obtained by their superposition. If the modified intensity gets to a maximum, it is said that they have interfered constructively, or that a constructive interference has occurred. If the resultant or modified intensity is zero (minimum), it is termed as a destructive interference (see Fig. 6.1).

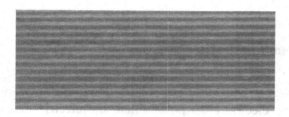

Fig. 6.1. A typical interference pattern representing a two beam interference.

6.2.1 *Conditions for the interference of light*

A stationary interference pattern is obtained when the following conditions are satisfied,

a) The two interfering waves are coherent.
b) The two interfering waves propagate in the same direction.
c) The two interfering waves are vibrating or have the components vibrating in the same plane (it implies that the state of polarisation of the two interfering beams is the same).

6.2.2 *Principles of interferometry*

a) Optical path difference (OPD) between a test beam and a reference beam will cause a difference in phase.
b) OPD should not be greater than the coherence length of the source beam.
c) Interference between these two beams will be constructive at some points and destructive at others, forming an interference pattern.

If I_1 and I_2 represent the intensity of the two interfering beams and $(\alpha_1 - \alpha_2)$ is the difference in phase between the phases of the two beams, then the universal interference equation can be written as,

$$I = I_1 + I_2 + 2\sqrt{I_1 I_2} \cos(\alpha_1 - \alpha_2) \qquad (1)$$

6.3 Two-Beam Interference

Two plane waves with different polarisations will interfere differently. For transverse electric (TE) polarisation, the electric fields of the two vectors overlap completely, whereas for other polarisation combinations, the extent of overlap depends on the angle between the two waves. First let us consider an interference pattern generated by two waves originating from a common coherent source and converging in the x-z plane at equal angle $'\theta'$ with respect to the z axis as shown in Fig. 6.2

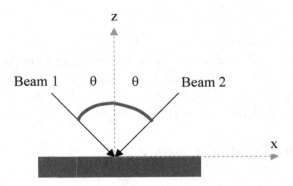

Fig. 6.2 Two-beam interference in x-z plane.

A one dimensional periodic feature can be fabricated by recording the interference of two coherent waves. The pattern representing the interference varies along the axis periodically in terms of the intensity with a periodicity, p, given by,

$$p = \lambda/2 \sin \theta \tag{2}$$

λ is the wavelength of the laser beam and θ is the half angle of the intersection of the two beams at the sample.

6.4 Laser Interference Lithography

Many applications require only a periodic or quasi-periodic pattern which could be achieved by a much simpler laboratory-scale technology, namely, interferometric lithography (IL). IL is a technique based on the interference of a number of (most often two or four) coherent laser beams, which produce patterns over large areas and volumes. The fundamental concepts of IL could be understood by considering the simple case of a two-beam interference (See Fig. 6.3.), where coherent laser beams are symmetrically incident from both sides. The period (p) of the interference pattern is given by $\lambda/2\sin\theta$. Standing wave patterns exist throughout the overlap between the beams as long as this overlap distance is shorter than the longitudinal coherence length of the laser beams and the wafer can be placed anywhere inside this coherence volume.

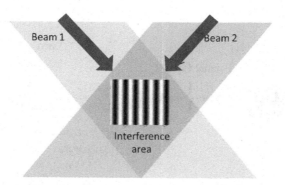

Fig. 6.3. A schematic representation of two-beam interference [18].

IL is a conceptually simple process, which uses a small number of coherent optical beams which are incident from different directions on a photosensitive layer to produce an interference pattern whose intensity distribution is recorded in the photosensitive layer and is later transferred based on thermal and/or chemical processes. The spatial-period of the features fabricated can reach a minimum of half the wavelength of the interfering light [18]. These techniques of IL, which bears different names; such as holographic lithography and mask less laser interference photolithography, can be divided into amplitude division and wave front division lithography (See Figs. 6.4 and 6.5) [19–21].

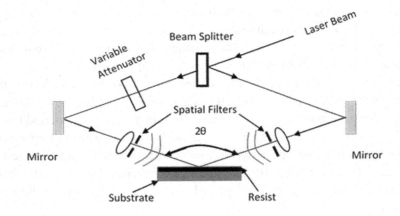

Fig. 6.4. Conventional configuration for interference lithography [19].

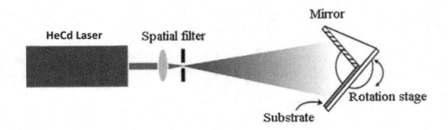

Fig. 6.5. Lloyd's mirror configuration for interference lithography [19].

The resolution of optical lithography system is usually expressed in terms of its wavelength (λ), a constant, depending on the lithographic process condition, and the numerical aperture (NA) as,

$$\text{Resolution} = \frac{\lambda K}{NA} \tag{3}$$

In IC manufacturing employing mask based photolithography, typically, this optics and the wafer stage are either placed in air or in a vacuum, to reduce the light absorption by the surrounding medium. This means that the refractive index (n) is unity and $NA = \sin \theta$. The numerical aperture therefore cannot be greater than 1, and more practically, takes on values in the range from 0.5 to 0.8, with a higher number reflecting a less stringent process. The NA of optical lithography tools ranges from about 0.5 to 0.6. Immersion technology provides another simple way to increase numerical aperture: by making $n > 1$. This could be achieved by using liquid (all liquids have a refractive index greater than unity) between the projection optics and writing medium as shown in Fig. 6.6. [22, 23]. Typically, the commercially-available index matching liquid used has a refractive index ranging from 1 to 1.6. It should be noted that the high index liquids should have sufficient transmission at the desired wavelength, which restricts the increase in index far greater than 2.

Generally, the concept of classical optical interference lithographic techniques is diffraction limited and features that can be written have a minimum size limit defined by $\lambda/2$, where λ is the optical wavelength [24, 25]. For a two-beam interference in air or vacuum, the period (p) of the pattern is given by equation described as (2) before.

Using an immersion concept, employing a liquid of index $n > 1$, the period can be reduced further as per the below formulation,

$$p = \frac{\lambda}{2 \, n \sin \theta} \tag{4}$$

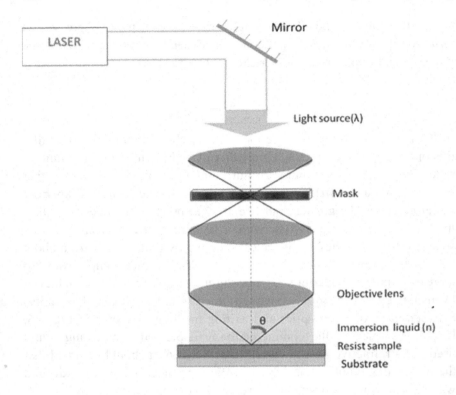

Fig. 6.6. Schematic diagram of mask based projection lithography system using an immersion liquid.

But the feature size can be reduced further by using the concept of immersion lithography where the refractive index of the surrounding medium is changed to increase the numerical aperture with respect to $NA = n \sin \theta$. For example, if θ is fixed at 45°, and purified water (n=1.33) is employed as a liquid medium, NA increases from 0.707 to 1.01. Moreover, if a liquid of $n = 1.8$ is employed, NA becomes 1.27. Various configurations have been investigated to date to fabricate features using immersion interference lithography [18,20,26,27]. Most of the reported work explores the implementation of the immersion concept using a deep ultraviolet (DUV) light source [20,23,26]. In one such system developed by IBM research, a 193 nm light enters the vacuumed interferometric body from the top and is split into two beams by diffraction from a fused silica grating [28]. These beams overlap at the wafer surface after reflecting from two mirrors and pass through a custom fused-silica prism. Even though this configuration provides better resolution, it requires deep UV optics for its implementation, which is very expensive. Moreover, most configurations require a larger number of optical components like beam splitters, mirrors, etc.

6.5 Four-Beam Interference

The polarisation states of the beams involved in the interference generally play a major role in the image contrast. So the major effect of polarisation in cases of the interference between two beams, is that it varies the image contrast. In the case of a multiple beam interference, such as a four-beam interference, the situation is more complicated and needs to be handled properly. Two-dimensional periodic intensity distributions with a square symmetry can be formed by interfering four coherent plane waves that converge at equal angles along two perpendicular planes. A good configuration to realise such a four beam interference configuration is shown in Fig. 6.7.

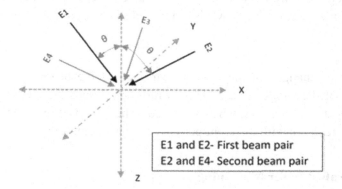

Fig. 6.7. Four-beam interference configuration.

6.6 Fabrication System and Methodology

The experimental set up for four-beam and five-beam interference lithography is shown in the Fig. 6.8. A spatially filtered, 364 nm CW laser beam is directed onto an electronic shutter with a temporal resolution of 10 ms, which controls the exposure time. A quarter wave plate follows the shutter, which converts the laser beam into a circularly polarised beam. This circularly polarised beam is directed perpendicularly onto a custom-fabricated cross-grating (Ibsen Photonics Pte Ltd), which diffracts the input beam into multiple beams. The cross-phase grating is designed to generate four first-order beams and a zero-order beam. A linear polariser, mounted on a custom fabricated rotating mount, is placed in the optical path of each diffracted beam to obtain either an s-polarised or p-polarised beam. An optional hard stop is employed to block the zero-order beam so as to realise a four-beam interference. The diffracted beams are directed such that they are about 45° with respect to the z-axis and they are made to fall on the resist, each at an angle of 35° with respect to the z-axis. The substrate used is a silicon wafer, coated with a positive resist, AZ 9260. A 1 μm thick layer of resist is coated using a spin coater at a speed of 4000 rpm for 30 sec. After exposure, the features patterned are developed in an AZ400K developer, followed by a DI water rinse. Field emission scanning electron microscopy (FESEM) is employed to characterise the developed patterns.

In general, changes in the polarisation state of the beams changes the contrast of the fringe pattern. This in turn determines the contrast of the photoresist line features when recorded. The polarisation changes also result in features with different periodicity and orientation.

6.7 Ultrafast Laser Machining

Various lasers with different pulse durations are nowadays being used in 3D printing and manufacturing. The ultrashort pulse lasers enable higher resolution for laser machining compared to conventional lasers [24–26].

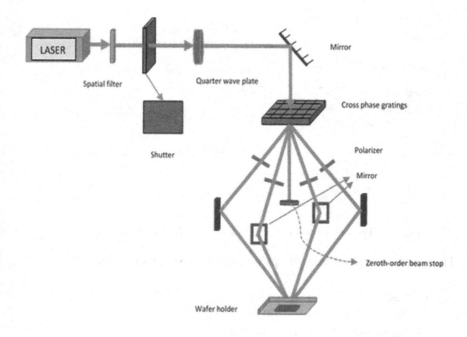

Fig. 6.8. Experimental setup for grating based multiple-beam interference lithography.

This section describes and compares the physical phenomena which take place when the laser pulses with different durations interact with the material surface.

6.7.1 *Ablation mechanisms depending on laser pulse duration*

There are three different time regimes for photon energy delivery: slow, medium and fast. These rates can be categorised by the physical time scale in which the interaction between the laser pulses and the target material occurs as shown in Fig. 6.9. When laser pulses are irradiated to a target material, carrier excitation, thermalisation and thermal/structural events sequentially take place. If the pulse duration is relatively long (e.g. longer than ms), all three phenomena occur. As the pulse duration gets shorter, thermal and structural effects lessen [27–29].

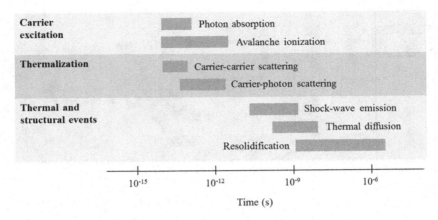

Fig. 6.9. The interaction of laser pulses with solids at different time scales.

The time scales used for laser pulse durations are given below:

- Second   = 1 s
- Millisecond   = 1/1,000 s
- Microsecond   = 1/1,000,000 s
- Nanosecond   = 1/1,000,000,000 s
- Picosecond   = 1/1,000,000,000,000 s
- Femtosecond   = 1/1,000,000,000,000,000 s
- Attosecond   = 1/1,000,000,000,000,000,000 s
- Zeptosecond   = 1/1,000,000,000,000,000,000,000 s

The time scale longer than a nanosecond (ns) is regarded to be in the slow regime; the medium regime comprises pulse durations between a picosecond (ps) and a nanosecond (ns); the fast regime is a pulse duration shorter than a picosecond. There are various kinds of Continuous-Wave (CW) lasers, e.g. diode laser, HeNe laser, CO_2 laser and Nd:YAG laser. They can be basically operated in a pulsed mode by modulating pump sources or by installing external intensity modulators. The attainable pulse duration is down to a nanosecond at best. For shorter pulses, shorter than a nanosecond, intra-cavity loss or intra-cavity gain should be modulated; Q-switching is a representative technique for this purpose. Ultrashort pulse lasers having a pulse duration less than 1 ps require a dedicated phase-alignment scheme, which is called mode-locking. Using various mode-locking schemes, e.g. Kerr lens mode-locking, nonlinear polarisation rotation and saturable absorption, ultrafast femtosecond laser pulses can be generated and be applied to high resolution nano/micro 3D printing and manufacturing. For ultrashort pulse generation, special gain materials supporting broadband lasing, is required; examples are, Ti:Sapphire crystal, Cr:Forsterite crystal and rare-earth-doped optical fibres (Er, Yb, Tm and Ho fibre).

'Ablation' is the removal of material by the direct absorption of laser energy. By selecting the laser pulse duration, one can determine which physical phenomena is to be incorporated within the 3D printing and manufacturing process as shown in Table 6.1. Ablation processes are classified into two types: photothermal and photochemical processes depending on the incident pulse's duration. If the pulse is longer than the thermal relaxation time, the photothermal process is dominant; otherwise, the photochemical process is. With ultrashort pulses, Coulomb explosion will dominate the material ablation. In this case, the laser energy is confined within a small volume (area) because the pulse duration is too short, so the pulse energy cannot be efficiently dissipated to its surroundings by heat conduction, convection or radiation. This energy confinement effect with ultrashort laser pulses enables high-resolution manufacturing without unexpected thermal effects to the surrounding area.

Photothermal process
- Mechanism includes thermal stresses generation, evaporation and sublimation.
- More volatile species are removed first, which results in change in chemical composition.
- At high fluences, boiling results in a phase explosion.

Photochemical process
- Direct ionisation of the molecules.
- Results in a thermal phase transformation, direct bond-breaking and explosive lattice disintegration via electronic repulsion.

- Direct bond-breaking is possible in polymers, ablation regions and small heat affected zones (HAZs)

Table 6.1. Different physical phenomena with different time scales.

Process (Time)	Physical Phenomena	Material Change
Slow	Thermal heating Photo-thermal effects Thermal stress	Evaporation Spallation
Fast	Photochemical Photophysical	Evaporation
Ultrafast	Coulomb explosion Mechanical shock	Evaporation Coulomb explosion

6.7.2 Long pulse laser processing

When a CW or long pulsed laser beam reaches a target material, linear optical absorption is the dominating process. If the target material has a bandgap of E_g, the photon can be absorbed when the incident photon has a high energy, enough to fill the bandgap ($h\nu > E_g$) (see Fig. 6.10). If the photon energy is less than the bandgap level, energy cannot be absorbed. In the linear absorption regime, the photon energy is a function of the wavelength, so the wavelength of the input beam determines the overall process efficiency. Therefore, the laser wavelength needs to be carefully selected considering the absorption wavelength spectrum of the material. The absorbed energy would be transferred to the surrounding area by thermal conduction, convection and/or radiation. The excessive heat generates unexpected effects, such as HAZ, micro-cracks and debris, to the target material.

Long pulse ablation (longer than ns)
- Because of the long pulse duration, thermal equilibrium between the electrons and the lattice is made lesser than a single pulse duration.

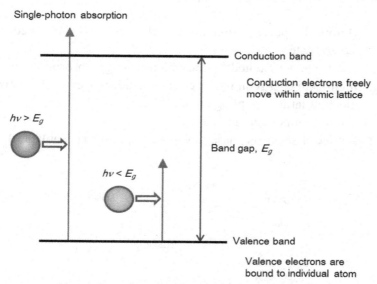

Fig. 6.10. The photon energy absorption process in the linear absorption regime.

- Ablation occurs by melting and vaporisation. These material's phase changes are due to the temperature rise.
- Excessive heat leaves behind HAZ, micro-cracks, surfaces debris and so on (see Fig. 4.12)
- Traditional laser machining processes, such as laser cutting, laser drilling, laser welding, laser heat treatment and laser cladding, are based on in this regime, where heating is dominant.

6.7.3 *Short pulse laser processing*

When short laser pulses, with a pulse duration between a picosecond and a nanosecond, reach the target material, short pulse laser processing occurs. There are much more dynamic thermal interactions between the electrons and the lattice in this regime, which can be explained by the two-temperature model (see Fig. 6.11). Due to the heat diffusion effect, HAZ still exists, but in a lesser amount than that produced by the long pulse regime.

Two temperature model (see Fig. 6.11)

(1) Before the pulse irradiation, the electrons and the lattice are in steady state.

(2) Electrons get excited by the photon energy absorption and then high-level energy in the electrons are transferred to the lattice by election-lattice coupling effect.

(3) Lattice temperature increases.

(4) Lattice changes its phase to vapour or plasma (i.e. ablation).

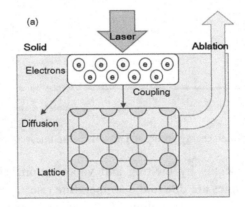

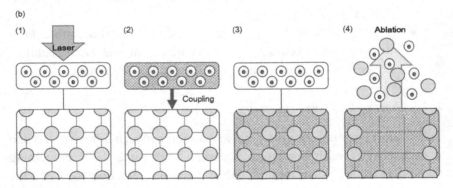

Fig. 6.11. Two temperature model for short pulse laser processing. (a) Definitions and basic model; (b) laser-material interaction process: (1) before the pulse's arrival, (2) electrons get excited, (3) the energy is transferred to the lattice via electron-lattice coupling and (4) short pulse ablation occurs.

6.7.4 *Ultrashort pulse laser processing*

When ultrashort laser pulses (less than 1 ps) are focused onto the target material, ultrashort pulse laser processing occurs. Due to the very short pulse duration, no heat transfer occurs between the electrons and the lattice. Therefore, the photon energy is confined within the electrons and cannot be disspated to the surroundings. If the energy exceeds a certain threshold, the electrons escape from the material by ionization, which is explained by the Coulomb explosion. This process does not include hydrodynamic motions or fluid dynamics; it is based on diect solid–vapour transition. Thus, non-thermal ablation is possible with negligible heat-induced side effects, such as HAZ, micro-cracks and debris to the material (see Fig. 4.13).

Other than the heat transfer, another important mechanism in ultrashort pulse laser processing is nonlinear absorption. Unlike linear absorption, nonlinear absorption is a function of both the wavelength and the peak intensity of the incident laser beam. Therefore, in the nonlinear absorption regime, even with a lower photon energy (which corresponds to a longer wavelength) than the material bandgap, the photon energy can be absorbed by the material (see Fig. 6.12). Therefore, any materials can be processed using ultrashort pulses regardless of the bandgap; this cannot be realised in conventional laser machining.

The advantages and potential applications of ultrafast laser processing are summarised below:

Advantages of ultrafast laser processing
- High resolution
- Non-thermal process
- HAZ suppression
- Absence of plasma shielding
- Micro and nano structuring
- Two-photon polymerisation
- Multiphoton absorption
- Internal structural modification.

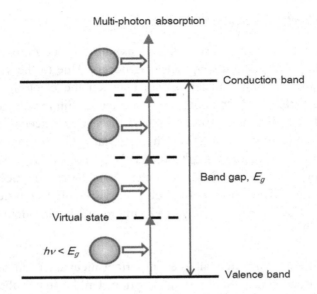

Fig. 6.12. The laser–material interaction in the case of using an ultrashort pulse laser.

Applications of ultrafast laser processing

- Direct laser lithography:
  High resolution patterning can be realised without the need for masks or repetitive chemical etching processes.
- Wafer dicing (stealth dicing):
  Traditional laser dicing for semiconductors, display products and solar panels suffer from residual thermal effects, e.g. surface contamination, HAZ, debris and property change. Ultrafast laser machining enables very clean cuts without thermal issues.
- Via hole drilling:
  Semiconductors and PCB industries are moving towards multi-layer structures by stacking wafers or circuit boards. Interlayer connection is one of the critical issues, which can be resolved with a high-resolution ultrafast laser drilling.
- Fabricating flat-panel displays (FPD):
  Defects inside the FPD cells can be repaired using ultrafast laser machining. The importance of non-thermal ultrafast machining is

rising because organic materials, such as the Organic LED (OLED), are very susceptible to heat energy.

- Fabrication of microfluidic structures:
  Due to the nonlinear absorption, the microfluidic inner channels with micrometre dimensions can be fabricated directly instead of through a repetitive lithography process.

- Fabrication of optical waveguide:
  An ultrafast laser can change the refractive index of transparent materials that can guide the optical wave inside the high index region like optical fibres.

6.8 Summary

The basics of interferometry, two-beam, four-beam and five-beam interference based patterning in 2D and 3D are detailed in this chapter from a fundamental and application point of view. A detailed analysis of the intensity distributions for various cases of the four-beam interference (where the beams are converged at equal angles in two orthogonal planes) has been performed for different polarisation combinations. It should be mentioned that the concepts detailed in this chapter can be adopted by students to develop new variants for realising different types of 2D and 3D structures on various substrates, taking into consideration the correct wavelength and material combination. Further, one can even explore the options to realise different crystal lattice structures with a laser interference writing approach. The extensive references outlined in (and listed at the end of) this chapter can serve as an excellent resource for students at the master and research level.

Problems

1. With the help of a schematic diagram, explain briefly the working principle of a two-beam laser interferometer.

2. Write the different conditions to be met for obtaining interference between waves.

3. What are the principles of interferometry?

4. Briefly explain how patterning resolution is related to light wavelength and numerical aperture in an optical lithography system.

5. Describe the working principle of a mask-based immersion lithography system with a schematic diagram.

6. Discuss the effect of the polarisation of the light beams in patterning using laser interference.

7. Explain the role of phase grating in the multiple beam interference system used for patterning.

8. Sketch the optical configuration of multiple beam interference patterning system and mark its components.

9. List the differences between photothermal and photochemical processes.

10. Explain the heat effects in CW or long-pulse laser machining with a clearly labelled drawing.

11. Explain the short pulse laser processing based on the two temperature model.

12. List five advantages of ultrafast laser processing.

13. List five applications of ultrafast laser processing.

References:

[1]   J. P. Silverman, Challenges and progress in x-ray lithography, (Silverman, 1998) *Journal of Vacuum Science and Technology* B **16**, 3137–3141 (1998).

[2]   E. Toyota, and M. Washio, *Extendibility of proximity x-ray lithography to 25 nm and below*, (AVS, Anaheim, California (USA), 2002), pp. 2979–2983.

[3]   A. Tseng, A. Notargiacomo, and T. P. Chen, Nanofabrication by scanning probe microscope lithography: A review, *Journal of Vacuum Science & Technology B: Microelectronics and Nanometer Structures* **23**, 877–894 (2005).

[4] D. S. Ginger, H. Zhang, and C. A. Mirkin, *The Evolution of Dip-Pen Nanolithography*, Angewandte Chemie International Edition **43**, 30–45 (2004).

[5] R. D. Piner, J. Zhu, F. Xu, S. Hong, and C. A. Mirkin, *"Dip-Pen" Nanolithography*, Science **283**, 661–663 (1999).

[6] S.R. J. Brueck. Optical and Interferometric Lithography — Nanotechnology Enablers, *Proceedings of the IEEE*, 93, 1704–1721 (2005)

[7] R. C. F. J.M. Carter, T.A. Savas, M.E. Walsh, and T.B. O'Reilly (M.L. Schattenburg and H.I. Smith), *Interference Lithography, in Submicron and Nanometer Structures,* (MTL Annual Report, 2003), 186–188.

[8] J. de Boor, D. S. Kim, and V. Schmidt, Sub-50 nm patterning by immersion interference lithography using a Littrow prism as a Lloyd's interferometer, *Opt. Lett.* **35**, 3450–3452 (2010).

[9] Byun, and J. Kim, Cost-effective laser interference lithography using a 405 nm AlInGaN semiconductor laser, *Journal of Micromechanics and Microengineering* **20**, 055024 (2010).

[10] S. Owa, and H. Nagasaka, Advantage and feasibility of immersion lithography, *Journal of Microlithography, Microfabrication, and Microsystems* **3**, 97–103 (2004).

[11] Hoffnagle, Liquid immersion deep-ultraviolet interferometric lithography, *J. Vac. Sci. Technol. B 17*, 3306 (1999).

[12] A.N. Boto, P. Kok, D. S. Abrams, S. L. Braunstein, C. P. Williams, and J. P. Dowling, Quantum Interferometric Optical Lithography: Exploiting Entanglement to Beat the Diffraction Limit, *Physical Review Letters* **85**, 2733–2736 (2000).

[13] S. Okazaki, Resolution limits of optical lithography, *J. Vac. Sci. Technol. B* **9**, 2829–2833 (1991).

[14] T. M. Bloomstein, M. F. Marchant, S. Deneault, D. E. Hardy, and M. Rothschild, 22-nm immersion interference lithography, *Opt. Express* **14**, 6434–6443 (2006).

[15] R. Dammel, F. M. Houlihan, R. Sakamuri, D. Rentkiewicz, and A. Romano, 193 nm Immersion Lithography — Taking The Plunge, *Journal of Photopolymer Science and Technology* **17**, 587–601 (2004).

[16] H. C. Pfeiffer, R. S. Dhaliwal, S. D. Golladay, S. K. Doran, and M. S. Gordon, Projection reduction exposure with variable axis immersion lenses: Next generation lithography, *J. Vac. Sci. Technol. B* **17**, 2840–2846 (1999).

[17] Nanophotonics: Accessibility and Applicability by Committee on Nanophotonics Accessibility and Applicability, pp1, (*The National Academies Press, Washington DC,2008*).

[18]   R. Sidharthan, F. Chollet, V.M. Murukeshan, Periodic patterning using multi-facet prism based laser interference lithography, *Laser Physics*, **19** (3), 505–510, (2009).

[19]   J. Chua and V. Murukeshan, Patterning of two-dimensional nanoscale features using grating-based multiple beams interference lithography, *Physica Scripta* **80**(1), 015401 (2009).

[20]   J. K. Chua and V. M. Murukeshan, UV laser-assisted multiple evanescent waves lithography for near-field nanopatterning, *Micro & Nano Letters* **4**(4), 210–214 (2009).

[21]   J. K. Chua, V. M. Murukeshan, S. K. Tan, Q. Y. Lin, Four beams evanescent waves interference lithography for patterning of two dimensional features, *Optics Express* **15**(6), 3437–3451 (2007).

[22]   R. Sidharthan. & V. M. Murukeshan. Nano-scale patterning using pyramidal prism based wavefront interference lithography, *Physics Procedia* **19**, 416–421 (2011).

[23]   R. Sidharthan., V. M. Murukeshan., Pattern definition employing prism-based deep ultraviolet lithography, *Micro and Nano Letters* **6**(3), 109–112 (2011).

[24]   Meijer, J., Du, K., Gillner, A., Hoffmann, D., Kovalenko, V. S., Masuzawa, T., ... & Schulz, W. (2002). *Laser machining by short and ultrashort pulses, state of the art and new opportunities in the age of the photons*. CIRP Annals-Manufacturing Technology, **51**(2), 531–550.

[25]   Cerami, L., Mazur, E., Nolte, S., & Schaffer, C. B. (2013). *Femtosecond laser micromachining. In Ultrafast nonlinear optics* (pp. 287–321). Springer International Publishing.

[26]   Gattass, R. R., & Mazur, E. (2008). *Femtosecond laser micromachining in transparent materials*. Nature Photonics, **2**(4), 219–225.

[27]   Della Valle, G., Osellame, R., & Laporta, P. (2008). *Micromachining of photonic devices by femtosecond laser pulses*. Journal of Optics A: Pure and Applied Optics, **11**(1), 013001.

[28]   Liu, X., Du, D., & Mourou, G. (1997). *Laser ablation and micromachining with ultrashort laser pulses*. IEEE Journal of Quantum Electronics, **33**(10), 1706–1716.

[29]   Perry, M. D., Stuart, B. C., Banks, P. S., Feit, M. D., Yanovsky, V., & Rubenchik, A. M. (1999). *Ultrashort-pulse laser machining of dielectric materials*. Journal of Applied Physics, **85**(9), 6803–6810.

Chapter 7

LASER SAFETY AND HAZARDS

Laser-based 3D printing and manufacturing utilises high power lasers whose power exceeds certain thresholds required to change the status of materials, such as melting, evaporation or a direct change to the plasma state. These high power lasers are very hazardous; a direct or indirect exposure can burn the retina of the eye or the skin. In order to control the risk of laser related injury, the sale and usage of lasers are subject to government regulations depending on the power and wavelength. These regulations prescribe how to label lasers with specific warnings and how to select proper safety goggles when operating the lasers. This chapter covers the topics of safety design and the proper use and implementation of lasers for minimal risks and hazards.

7.1. Introduction

Laser is a coherent light source in both time and spatial domains [1,2]. High coherence in the time domain implies that the optical phase of a laser is well aligned along the wavefront so that a constructive interference can be easily attained. Thanks to its high coherence, mode-locked ultra-short pulses, even attosecond (10^{-18} s) pulses, with high peak power can be generated by a laser, which is not the case with traditional white-light sources [3–5]. High coherence of the laser in the spatial domain provides high directionality and low divergence of light with a negligible loss over long distances. For a future space mission (LISA: Laser Interferometry Space Antenna Project), a well-designed telescope system is being designed to transmit the laser beam over a 5,000,000 km, to detect gravitational-wave induced strains in space-time. High coherence of the laser light, therefore, is beneficial for diverse laser applications, including 3D printing. However, the high coherence also raises important safety issues if not properly designed and implemented. In the time domain, the high peak power of an ultra-short pulse can reach up to terawatt (TW, 10^{12} W), which corresponds to the instantaneous world power consumption. In the spatial domain, the laser light can be focused on a small spot, less than 500 nm in diameter. If an intense, pulsed laser light is focused on (most) metals or even transparent glass, the material can be directly melted into the liquid state or be ablated into the plasma state. A human eye or skin will be no exception. Therefore, in order to prevent the possible exposure to laser hazards, this chapter will provide information on basic laser hazards, safety standards, safety goggles and other precautions for laser-assisted 3D printing [6–14].

7.2. Basic Hazards from Lasers

Lasers can cause damage in biological tissues, both to the eye and to the skin, based on two representative mechanisms: photo-thermal and photo-chemical interaction, as shown in Fig. 7.1 [15–19]. Photo-thermal damage is the predominant cause of injury. When the eye or skin is exposed to a laser radiation, the tissues are heated up to the point where proteins can be denaturised. Even moderate power lasers can injure the retina inside the eye by photo-thermal effects due to its focusing geometry.

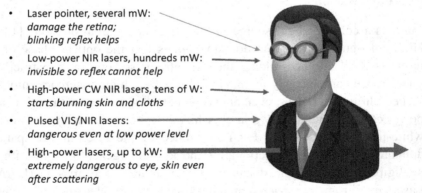

- Laser pointer, several mW: *damage the retina; blinking reflex helps*
- Low-power NIR lasers, hundreds mW: *invisible so reflex cannot help*
- High-power CW NIR lasers, tens of W: *starts burning skin and cloths*
- Pulsed VIS/NIR lasers: *dangerous even at low power level*
- High-power lasers, up to kW: *extremely dangerous to eye, skin even after scattering*

Fig. 7.1. Laser safety issues depending on incident laser power level.

Photo-chemical damage is caused by the chemical reactions in tissue when exposed to and triggered by the laser light. Due to its high photon energy, short-wavelength light (e.g. blue or ultra-violet) can easily generate photo-chemical damages. Wavelength-dependent representative pathological effects to eye and skin are listed in Table 7.1 [20,21]. Ultra-short near-infrared laser pulses can also cause photo-chemical damages through the multi-photon absorption process. There are other sources of laser hazards, such as shock wave by explosive boiling, electric shock or the unexpected breathing of vaporised materials.

Table. 7.1. Laser safety issues depending on incident laser power level.

Wavelength range	Pathological effect	
	Eye	Skin
200–280 nm (UV-C)	Photokeratitis (inflammation of the cornea, equivalent to sunburn)	Erythema (Sunburn) Skin cancer Accelerated skin aging
280–315 nm (UV-B)	Photokeratitis	Increased pigmentation
315–400 nm (UV-A)	Photo-chemical cataract (clouding of the eye lens)	Pigment darkening Skin burn
400–780 nm (Visible)	Photo-chemical damage to the retina and thermal retina injury	Pigment darkening Photosensitive reactions Skin burn
780–1400 nm (IR-A)	Cataract and retinal burn	Skin burn
1.4–3.0μm (IR-B)	Corneal burn, aqueous flare, (protein in the aqueous humour), cataract	Skin burn
3.0–1000 μm (IR-C)	Corneal burn only	Skin burn

The spectral sensitivity and intensity response of the human eye to the laser light is an important issue in laser safety. Keeping oneself safe from possible laser hazards is not a simple task because our eyes cannot detect ultraviolet (UV) wavelengths shorter than 380 nm, and infrared (IR) wavelengths longer than 800 nm. Under daylight condition, the average normal-sighted human eye is most sensitive at a wavelength of 555 nm, resulting in the fact that green light at 555 nm procures the impression of highest brightness compared to light at other wavelengths, as shown in Fig. 7.2. Because the spectral sensitivity of the human eye at 490 nm is about 20% of its sensitivity at 555 nm, if lasers at 490 nm and 555 nm look to provide the same intensity to the human eye, the power at 490 nm is 5 times stronger than that at 555 nm. There are three representative laser wavelengths widely used in 3D printing: UV lasers (e.g. 355 nm) for polymer, metal and ceramic materials, visible lasers (e.g. 532 nm) for metals and IR lasers (10.6 μm) for polymers. However, to our eyes, conventional CCD or CMOS cameras (designed to have similar spectral sensitivity with the human eye) cannot detect the UV or IR lasers, which is why we observes blue or green laser lights in 3D printers, even though the IR power is much stronger than the visible output. Therefore, for laser safety in 3D printing, the spectral response of the human eye, cameras and photo-detectors should be carefully considered in the safety design process. An IR/UV visualisation phosphor camera or viewing cards will help the first installation of the beam paths and isolation stage.

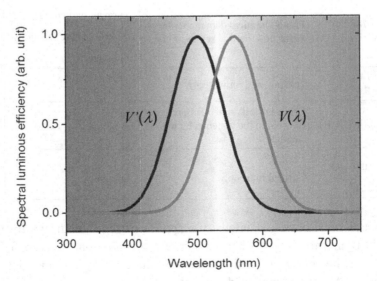

Fig. 7.2. Spectral luminous efficiency functions, $V(\lambda)$ for photopic vision and $V'(\lambda)$ for scotopic vision.

7.2.1. *Beam-related hazards*

Direct exposure to the laser beam can cause eye and skin damages. Laser beams are hazardous because they maintain their high intensities even after propagation over long distances owing to their high spatial coherence. Even when the intensity at the entrance of the eye is at a moderate level, the laser beam is focused on a small spot on the retina, where it can also cause serious permanent damage within a very short period of time. The skin is usually much less sensitive to laser than the eye, but an excessive exposure to ultraviolet light can cause short- and long-term photo-chemical damages similar to sunburn, while visible and infrared wavelengths are mainly harmful due to the thermal damage mechanism.

Eye damage

If a laser beam enters the eye by accident, they will pass through the cornea, aqueous humour, iris, lens and vitreous body, before finally being imaged on the retina as illustrated in Fig. 7.3. Because each part of the eye has a different absorptance, transmittance and reflectance depending on the wavelength, the injured area will also be dependent on the input wavelength. The cornea and the lens absorb UV light, so the damage is mainly located in the front segment of the eye. Because only a small portion of UV light reaches the retina, the relevant damage is minimal. Visible and NIR (IR-A) light can pass through the cornea, aqueous humour, lens and vitreous body with a low loss so they mostly affect the retina.

The critical damage mechanism in the retina is made by the focusing geometry of the eye. Because, the eye focuses visible and NIR light onto the retina, the laser beam can be focused onto a high-intensity spot on the retina which could be up to 200,000 times higher than at the point of the entrance of the eye. As the result, a moderately powered laser can cause serious permanent damage to the retina. Mid- and far-IR laser beams (IR-B and IR-C) show similar behaviour as UV ones; they are mostly

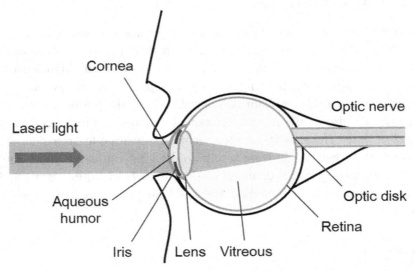

Fig. 7.3. Simplified cross-section of the human eye under laser irradiation: laser injury to the eye.

absorbed in the cornea and lens so they leave damage in the front of the eye; their effect on the retina is insignificant (see Fig. 7.4).

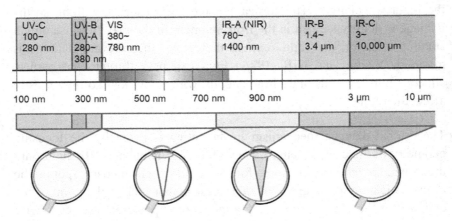

Fig. 7.4. Spectral transmission of light through the human eye to the retina.

UV is mainly absorbed in the front of the eye. The lens strongly absorbs UV light with wavelengths shorter than 400 nm, whereas the core absorbs wavelengths shorter than 300 nm, as shown in Figs. 7.4 and 7.5. The aqueous humour has a weak absorptance of less than 5% for wavelengths between 300 and 450 nm. In the UV range, the injury is dominated by photo-chemical damage. Even at a relatively low power level, e.g. with UV lamps, this photo-chemical damage can occur; this is because chemical reactions are triggered by UV light and is accumulative. Intense UV lasers can cause corneal flash burns as well as photokeratitis and cataracts in the eye's lens.

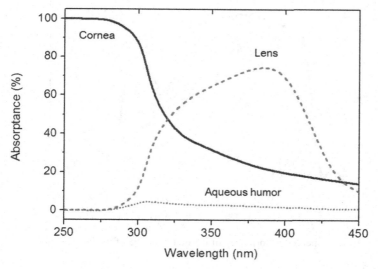

Fig. 7.5. Spectral absorptance of the frontal parts of the human eye (cornea, lens and aqueous humour).

Visible and NIR (IR-A) wavelengths between 400 and 1400 nm have high transmittance at the cornea, aqueous humour, lens and vitreous body (see Fig. 7.5). Thus, the laser beams in the visible and NIR range will penetrate the front part of the eye and the eyeball and be focused onto the retina and may cause the heating or ablation of the retina (see Fig. 7.6). The moderately powered laser can generate a transient temperature increase at the retina because the melanin pigments in the pigment epithelium near to the retina absorb the light and convert it into heat. A minor temperature increase of just 10°C can damage the retinal photo-receptor cells. When the laser has a strong average power or provides a high peak power in the pulsed mode, the retina can be damaged permanently within a short exposure time, within a fraction of a second. For visible and NIR wavelengths, the blinking motion of the eye helps to prevent the retinal damage when exposed to low powered lasers but does not work for high power or pulsed lasers.

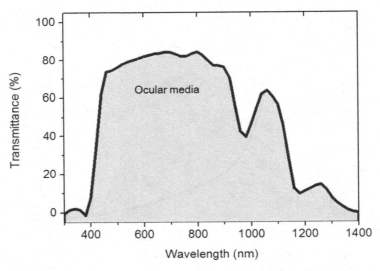

Fig. 7.6. Spectral absorptance of the ocular media of the human eye.

IR (IR-B and IR-C) lasers are particularly hazardous because our body's protective blink reflex does not respond to IR light. There are many different types of IR high power lasers widely used in 3D printing, e.g. Nd:YAG, Yb-fibre, Tm-fibre and CO_2 lasers, but the user may not recognise their exposure to the laser beam because there could be no pain or immediate damage to their eyesight; a minor click noise from the eyeball could be the only indication of retinal damage. Therefore, it is easy for users to be exposed to IR lasers for a long time. When the retina is heated over 100°C, a permanent blind spot is created by the local explosive boiling or direct ablation. There are some IR wavelengths around 3 μm and 10 μm with strong molecular absorption in the cornea, which could cause corneal injuries. Most of the damages caused by IR lasers are caused by the heating process; tissues are first heated and then the denaturation of proteins occurs.

If high power pulse lasers are involved, the materials' absorption characteristics differ drastically from those of moderately powered lasers. Although materials are transparent in a certain wavelength range, high-power laser pulses can be absorbed by most materials through the nonlinear absorption process. Therefore, visible and NIR lasers, originally transparent to the cornea, aqueous humour and lens, can be absorbed in the front part of the eye through nonlinear absorption. That is how laser-based eye surgeries and vision corrections are performed. Laser pulses shorter than 1 μs cause a rapid rise in temperature, resulting in the explosive boiling of water. The shock wave from the explosion can subsequently cause damage relatively far away from the point of impact. Therefore, the amount of light the eye can tolerate without damage is dependent on a number of circumstances: wavelength, intensity, pulse duration and so on. Thus, there are detailed rules for calculating the safe exposure limits (e.g. maximum permissible exposure, maximum permissible energy) for a given environment. Maximum permissible energy (MPE) calculation and selection rules for proper laser protective goggles will be separately covered in forthcoming sections.

Skin damage

Laser radiation can cause severe injury to the skin. Skin damage is considered less serious than eye damage because the functional loss of the eye is more debilitating. The damage threshold for both skin and eye are comparable except for the focused retinal region. For example, in IR (IR-B and IR-C) and UV wavelengths, where the beams are not focused on the retina, skin damage thresholds are similar to corneal damage thresholds, as shown in Fig. 7.7. When considering the skin's wider surface area, the possibility of skin damage is greater than that of eye exposure.

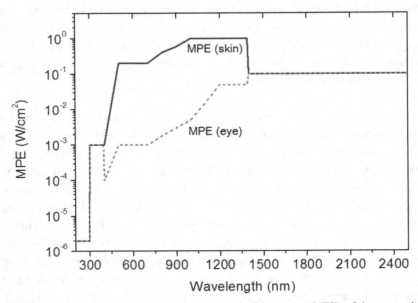

Fig. 7.7. Wavelength-dependent maximum permissible energy (MPE) of the eye and the skin (ANSI Z136.1-2007). In the UV and IR range, skin damage level is similar to that of the eye.

In laser skin hazards, three skin tissue layers are considered in general; epidermis, dermis and subcutaneous tissues. The epidermis is the outermost living layer of skin lying beneath the stratum corneum; the stratum corneum's thickness ranges from 10 to 20 μm. The thickness of the epidermis varies between 50 and 1.5 mm depending on the type of skin; the thinnest ones are on the eyelids and the thickest ones are on the soles. The epidermis serves as a barrier to protect the body from microbial pathogens, UV damage and chemical compounds. It also determines the skin colour depending on its melanin amount and distribution. As shown in Fig. 7.8, most visible and NIR (IR-A) light reach the epidermis after suffering from power attenuation while passing through the stratum corneum. Since IR (IR-B and IR-C) and UV are strongly absorbed by the stratum corneum, their effects on the epidermis is much lesser than those of the visible and NIR. The dermis is the skin layer between the epidermis and subcutaneous tissues. It consists of connective tissues—collagen, elastic fibres and extrafibrillar matrix—and cushions the body from external stress and strain. It also contains the blood vessels, the nerve cells, the mechano-receptors providing the sense of touch, and the thermo-receptors providing the sense of heat. The dermis also varies in thickness depending on its location; the thinnest ones are 0.3 mm on the eyelid and the thickest ones are 3.0 mm on the back. Most of the visible and NIR wavelengths that pass through the stratum corneum and the epidermis are strongly absorbed by the dermis (see Fig. 7.5). The subcutaneous tissue located beneath the dermis is the layer that stores fat. It contains blood vessels and nerves that larger than those in the dermis. The thickness of this layer varies throughout the body and from person to person. Some visible light unabsorbed by the dermis can reach the subcutaneous tissues.

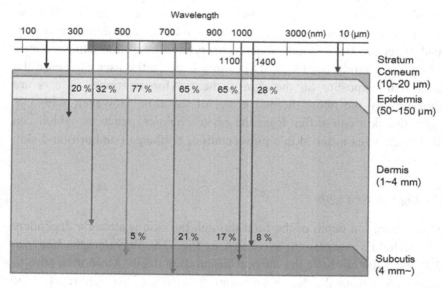

Fig. 7.8. Wavelength-dependent skin penetration of laser radiation.

The severity of laser damage depends on the following five factors: 1) exposure time, 2) light wavelength, 3) beam energy, 4) exposed area and 5) the type of tissue exposed to the beam. Laser damages are mostly related with the thermal effect; the laser beam firstly heats the tissues and then the denaturation of protein occurs as a result. If the laser beam has a short pulse duration, the explosive boiling of water could be the damage mechanism. The shock wave generated as the by-product of laser beam propagation can cause damages in regions relatively far from the impact points. The five factors that could cause laser damage will be dealt with in detail, one by one.

(1) *Exposure time*

When the skin is exposed to laser light in the UV, visible, NIR (IR-A) and IR (IR-B and IR-C) ranges for a short duration e.g. less than 10 seconds, the damage caused will largely be confined to the superficial layer, though it may involve the death of other layers of the skin (e.g. the stratum corneum and epidermis). This temporal injury can be painful depending on its severity but it will heal eventually; often without signs of injury. Burns made by a longer exposure over a larger area is more serious because the beam goes into the epidermis and dermis and may lead to serious loss of body fluids. This type of long-term exposure over a large area rarely occurs in 3D printing because the user can sense the laser light exposure on the skin as heat. Aforementioned effects are governed by the photo-thermal effect of the skin. However, with UV light, the skin can suffer from the photo-chemical reaction, which can lead to changes in the skin's pigmentation, erythema (sunburn) and skin cancer.

(2) *Light wavelength*

The penetration depth of the light through the skin is strongly dependent on the light's wavelength as shown in Fig. 7.5. Wavelengths between 700 and 1200 nm have the deepest penetration depth. More than 60% of such light reaches the dermis and more than 10% of the light reaches the subcutaneous tissues. Thus, this wavelength range is absorbed by the dermis, causing the deep heating of skin tissues. IR (IR-B and IR-C) light is absorbed by the superficial layer and induces the thermal effect; the resulting thermal skin burns by the light are similar to the burns made by other means. The exposure to visible, NIR and IR light causes an initial pigment-darkening effect, followed by erythema, if there is exposure to excessive levels. UV-A, UV-B and UV-C have different effects on the skin. UV-A (315 ~ 400 nm) can cause hyperpigmentation and erythema. UV-B and UV-C are absorbed by the epidermis layer and so cause erythema and blistering. Exposure to the UV-B range leaves the most serious injuries to the skin. In addition to thermal injury, UV-B (280~315 nm) may cause radiation carcinogenesis, either directly on the DNA or via the potential carcinogenic intracellular viruses. Exposure to UV-C (200~280 nm) and wavelengths longer than UV-A ranges are less harmful to human skin.

(3) *Beam energy*

High-power lasers also cause burns to the human skin. In 3D printing and manufacturing, various types of high-power lasers are used for melting polymers, metals and ceramics, either to initiate photo-polymerisation or for ablating the materials. On the basis of the laser's power, the light absorption can be split into two different ranges; linear and nonlinear absorption regimes. In most cases, light absorption is governed by the wavelength. However, if the power exceeds a certain threshold level, nonlinear absorption starts to be involved in the skin absorption mechanism. Therefore, power level also needs to be considered in addition to the wavelength. Short-pulse lasers (including ms, μs, ns, ps, fs and as lasers) have relatively higher peak power than continuous-wave lasers because the power is concentrated within the short pulse duration. Thus, the nonlinear absorption needs to be carefully considered when using short-pulse lasers.

(4) *Exposed area*

With the same laser energy, the injury will be severer and deeper if the size of exposed area is smaller. Since laser light maintains its beam size even after long distance propagation, users should be cautious even at long distances from the equipment. However, a smaller exposed area is beneficial in some cases, because one can easily sense the heat and realise one's exposure to laser light. The heat generated from the light absorption provides adequate warning to prevent severe thermal injury to the skin. Any irradiance of 0.1 W/cm^2 can be sensed at exposed skin diameters of larger than 10 mm. Lower irradiance of 0.01 W/cm^2 can be recognised if a large portion of the body is exposed. Long-term exposure to UV lasers has been shown to cause long-term delayed effects such as skin aging and skin cancer.

(5) *Type of tissue exposed to the beam*

Types of skin tissue—epidermis, dermis and subcutaneous tissues—have been covered in the previous section.

7.2.2. *Indirect non-beam hazards*

Although most laser hazards are due to the laser beam, there are other non-beam hazards that come from other units of the laser system, shown in Fig. 7.9. Some laser systems contain high electric voltage supplies, hazardous chemicals or exploding/imploding glass tubes [22–27]. Thus laser or laser-related equipment (including 3D printing or manufacturing equipment) designers, developers or users have also been injured by these non-beam hazards (see Figs. 7.10 and 7.11) [28,29].

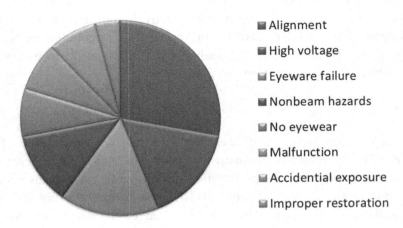

Fig. 7.9. Causes of laser accidents.

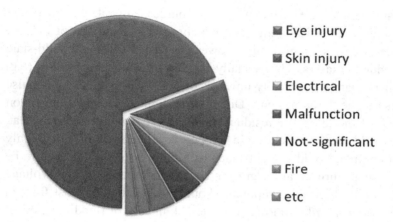

Fig. 7.10. Types of laser-related injuries.

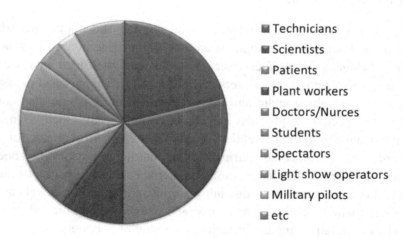

Fig. 7.11. Occupations of people that have incurred laser-related injuries.

(1) *Electric shock*

Lasers are pumped by different energy sources: optical, electrical and chemical ones. Electrical pumps are widely used as the direct or indirect secondary energy source for high-power gas (e.g. HeNe) and solid-state (semiconductor) lasers. They usually include high voltage devices (e.g. for discharge lamps); typically upwards of 400 V for a small 5 mJ pulse laser inside the laser system. Due to the limited energy conversion efficiency of the laser, the residual pump power is converted into heat, which should be removed by a proper cooling system. For highly efficient cooling, liquid coolants such as high-pressure water need to be used. These in turn create a greater hazard incorporating high-voltage electrical units. Electric equipment should be installed at least 0.25 m above the floor in order to lower electric risk in case of flooding. Optical tables, lasers and other equipment should be properly grounded. Enclosure interlocks should be respected and special precautions must be taken during troubleshooting.

(2) *Chemical hazards*

Chemical hazards include intrinsic gain, guiding and doping materials inside the laser system. Examples are beryllium oxide in argon ion laser tubes, halogens in excimer lasers, organic dyes dissolved in toxic or flammable solvents in dye lasers and heavy metal vapours and asbestos insulation in helium cadmium lasers. Chemical hazards can also come from unexpected by-products during laser processing; such as metal fumes from the cutting, welding or surface treatment of metals or the complex mix plasma made during the laser cutting of plastics. Poisonous fumes, dust or hot droplets of molten material can severely affect nearby users. For example, fumes containing arsenic, chromium or nickel can be produced when metals are machined and dangerous organic substances can be generated during the material processing of polymers.

(3) *Other hazards*

High temperatures and relevant fire hazards also need to be considered, especially during the operation of high-power lasers in Class IIIB or

Class IV. Secondary radiation (e.g. UV, EUV or even X-rays) can be generated when highly intense laser beams are focused onto certain targets and there is an extremely high rise in temperature. As this secondary radiation is not easy to consider in the design stage, careful precaution must be taken when handling extremely high-power lasers. Examples of mechanical hazards include moving parts in vacuum and pressure pumps; or the explosion or implosion of flash lamps, plasma tubes, water jackets and gas handling equipment. Cryogenic liquids, e.g. liquid nitrogen, can produce burns and replace oxygen in small unventilated spaces. Cryogenic liquids are explosive when ice forms in valves or connectors. When handling such cryogenic fluids, full protective equipment should be worn. Some laser chemicals are toxic, such as laser dyes and solvents; they are not only carcinogenic but can be flammable as well. There are still other contributing factors in laser safety as listed below.

- Unanticipated eye exposure during laser alignment
- Not using eye protection
- Incorrect eyewear selection
- Viewing laser generated plasmas
- No protection against ancillary hazards
- Inadequate training of laser personnel
- Failure to block beams or stray reflections
- Improper use of equipment
- Equipment malfunction
- Improper handling of high voltage
- Inhalation of laser generated air contaminants
- Lack of planning
- Poor housekeeping

7.3. Laser Ratings and Safety Standards

To give adequate guidance on proper handling methods and the necessary precautions to laser users, laser ratings are assigned to laser equipment; Class 1 is the least hazardous and Class 4 is the most hazardous. A laser's safety class is dependent not only on the laser's specifications (e.g. laser power, beam quality, wavelength and collimation) but also on the precaution level required (e.g. encapsulation). These classes can be used as a very simplified guideline because they provide the emission limits for only the laser unit without in-depth consideration of the beam delivery part, such as the beam diameter, the divergence, the focusing geometry and so on. Therefore, it is not a proper safety assessment of the full laser equipment. For the complete safety assessment of the laser equipment, e.g. a laser-based 3D printer, one should consider the details of the whole equipment and the printing process (i.e. how it is used).

7.3.1. *Laser ratings*

A laser's ratings also consider its visibility to our eyes and the operation mode [30,31]. Laser radiation includes broad wavelength spectra spanning over UV-C, UV-B, UV-A, visible, IR-A, IR-B and IR-C, as studied in Table 7.1. UV-A, visible light and IR-A are visible, whereas the others are invisible. The hazard level is also dependent on whether the laser output is a continuous wave or a short pulse. In a continuous wave (CW) laser, the average power (mW, W or kW) is used for its classification. On the other hand, in pulse lasers, the energy per pulse (nJ, μJ, mJ or J) or the peak power (kW, MW, GW or TW) is used for their classification. The laser classification system is used to indicate the level of laser hazard and maximum Accessible Emission Levels (AELs).

The previous classification system, which was based on five classes (1, 2, 3A, 3B and 4), has been replaced with a system of seven classes (1, 1M, 2, 2M, 3R, 3B and 4). It is an important duty of laser manufacturers and laser-related equipment manufacturers to classify their products and equip them with warning labels, safety key switches, interlocks and enclosure boxes. Users are to note that the classification may change when a user modifies a laser product; the user is then responsible for reclassification.

(1) *Class 1*

Class 1 lasers are safe under all conditions for normal use. These are safe under reasonably foreseeable conditions of operation, either because of the inherently low laser emission or because of its engineering design which implies that the laser system must be totally enclosed and human access to higher levels is not possible under normal operation. If the access panels of a totally enclosed system are removed for servicing or other purposes, the laser is no longer Class 1 and the precautions applicable to the embedded laser must be applied until the panels are replaced, for safety.

(2) *Class 1M*

Class 1M lasers are safe for all conditions of use except when passed through magnifying optics such as microscopes and telescopes. They emit wavelengths between 302.5 nm and 4000 nm, whose total output power is in excess of what is normally permitted for a Class 1 laser; but because of their diverging beams or very low power density, they do not pose a hazard in normal use and comply with the measurement conditions for a Class 1M product. They, however, may be hazardous to the eyes under certain conditions if magnifying optics are used with them.

(3) *Class 2*

Class 2 lasers are safe because the blink reflex limits the exposure time to less than 0.25 s. This reflex reaction provides adequate protection under reasonably foreseeable operating conditions e.g. the use of optical instruments. This class applies only to visible wavelengths between 400 and 700 nm. The average power is limited to 1 mW in the case of a continuous wave laser. Many laser pointers and lasers used in measuring instruments are in Class 2.

(4) *Class 2M*

Class 2M lasers are safe thanks to the blink reflex if not viewed through optical instruments. As with Class 1M, Class 2M applies to laser beams with a large diameter or a large divergence, for which the amount of light passing through the eye pupil cannot exceed the limits for Class 2. They, however, may be hazardous to the eyes under certain conditions if magnifying optics are used with them.

(5) *Class 3R*

Class 3R lasers are considered safe if handled with care and without direct intra-beam viewing. The average power of visible continuous lasers in Class 3R is limited up to 5 mW. The AEL is restricted to no more than five times the AEL of Class 2 for visible wavelengths and no more than five times the AEL of Class 1 for other wavelengths. The safety risk is lower than that of Class 3B lasers, so less manufacturing requirements and control measures apply to their users.

(6) *Class 3B*

Class 3B lasers are hazardous if directly exposed to the eye. Viewing the diffuse reflection is normally safe. The AEL for continuous lasers in the wavelength range from 315 nm to far infrared, is 500 mW. For pulse lasers of wavelengths between 400 and 700 nm, the limit is 30 mJ. Other limits apply to lasers with other wavelengths and to ultra-short pulsed lasers. Protective goggles are required for preventing the direct view of Class 3B lasers. Class 3B lasers must be equipped with a key switch and a safety interlock.

(7) *Class 4*

Class 4 lasers are the most hazardous lasers because Class 4 is the highest level; their use requires extreme caution. All high-power lasers whose power exceeds the AELs for class 3B are categorised in this class. Direct, indirect and diffuse beams can cause permanent eye damage and skin injuries. They may also be a fire hazard. These lasers must be equipped with a key switch and a safety interlock. Most industrial, military and medical lasers are in this category, including laser-based 3D printing and manufacturing systems.

You can refer Table 7.2 for information on the safety classes, laser types and examples.

Table 7.2. Laser safety classes, laser types and relevant examples.

Safety class	Laser type	Simplified description and examples
Class 1	Laser completely enclosed Very low power level	The accessible laser radiation is not dangerous whether power is less than a certain threshold or the radiation is under reasonable controlled environments. Examples: 0.2-mW laser diode; Enclosed 10 W Er-fibre or Yb-fibre lasers
Class 1M	Very low power level	The accessible laser radiation is not hazardous in the case of focusing optical elements not being used.
Class 2	Low power level Visible wavelength only	The accessible laser radiation is limited to the visible wavelength range (400–700 nm) with less than 1 mW average power. Thanks to the blink reflex, it is not dangerous for the eye in the case of limited exposure (up to 0.25 s). Prolonged staring may injure the eye, especially blue wavelengths. Examples: 1 mW HeNe laser for laser interferometry; Most laser pointers (not all)
Class 2M	Low power Visible Collimated large beam diameter or divergent	Same as Class 2, but with the additional restriction of no focusing optical element being involved. The power may be higher than 1 mW but the beam diameter should be large enough to limit the intensity to be safe for short-time exposure.

Table 7.2. (*Continued*).

Class 3R	Low power	The accessible radiation may be dangerous to the eye, but can have at most five times the permissible optical power of Class 2 (for visible radiation) or Class 1 (for other wavelengths). Examples: 5 mW three-mode HeNe laser
Class 3B	Medium power	The accessible radiation may be dangerous to the eye and, under special conditions, also for the skin. Diffuse radiation should normally be harmless. Up to 500 mW is permitted in the visible spectral region. Example: 100 mW continuous-wave frequency-doubled (532 nm) Nd:YAG laser
Class 4	High power	The accessible radiation is very dangerous for the eye and for the skin. Even light from diffuse reflections may be hazardous for the eye. The radiation may cause fires or explosions. Examples: 10 W argon ion laser; 4 kW thin-disk laser in a non-encapsulated setup

7.3.2. *Laser safety standards*

The laser ratings in the previous section are based on current European regulations. Since the 1970s, lasers have been classified into four classes depending on their wavelength and output power. The classification was further updated in 2002 and 2007. Since 2007, the classification system was revised to ones based on European IEC 60825 and ones based on US-oriented ANSI Laser Safety Standard (ANSI Z136.1). Important European and US laser safety regulations are listed below (see Figs. 7.12, Fig. 7.13 and Table 7.3).

European regulations

• IEC60825-1 – Safety of laser products
    Part 1: Equipment classification and requirements

- IEC60825-2 – Safety of laser products
   Part 2: Safety of optical fibre communication systems (OFCS)
- IEC60825-3 – Safety of laser products
   Part 3: Guidance for laser displays and shows
- IEC60825-4 – Safety of laser products
   Part 4: Laser guards
- IEC60825-5 – Safety of laser products
   Part 5: Manufacturer's checklist for IEC 60825-1
- IEC60825-6 – Safety of laser products
   Part 6: Safety of products with optical sources, exclusively used for visible information transmission to the human eye
- IEC60825-7 – Safety of laser products
   Part 7: Safety of products emitting infrared optical radiation, exclusively used for wireless "free air" data transmission and surveillance
- IEC60825-12 – Safety of laser products
   Part 12: Safety of free space optical communication systems used for transmission of information

- EN207 – Personal eye-protection equipment—Filters and eye-protectors against laser radiation (laser eye-protectors)
- EN208 – Personal eye-protection—Eye-protectors for adjustment work on lasers and laser systems (laser adjustment eye-protectors)

* IEC: International Electro-technical Commission (http://www.iec.ch)

US regulations

- ANSI Z136.1 – Safe use of lasers
- ANSI Z136.2 – Safe use of optical fiber communication systems utilizing laser diode and LED Sources
- ANSI Z136.3 – Safe use of lasers in health care
- ANSI Z136.4 – Recommended practice for laser safety measurements for hazard evaluation
- ANSI Z136.5 – Safe use of lasers in educational institutions
- ANSI Z136.6 – Safe use of lasers outdoors

Fig. 7.12. European style laser license sign.

- ANSI Z136.7 – Testing and labeling of laser protective equipment
- ANSI Z136.8 – Safe use of lasers in research, development or testing
- ANSI Z136.9 – Safe use of lasers in manufacturing environments

* ANSI: American National Standard Institute

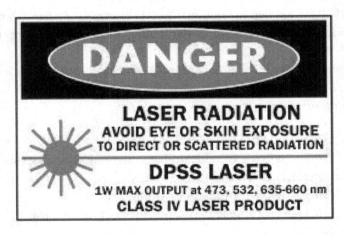

Fig. 7.13. US style laser license sign.

Table 7.3. Laser safety classes: comparison between IEC 60825, ANSI A136.1 and CDRH 1040.

IEC 60825	ANSI Z136.1	CDRH 1040	Safety Aspects
Class 1	1–1	I	Safe
Class 1M	-1M	IIa	Safe provided optical instruments are not used
Class 2	2	II	Visible lasers, safe for accidental exposures (< 25 ms)
Class 2M	-2M	IIIA	Visible lasers, safe for accidental exposures (< 25 ms) provided optical instrument are not used
Class 3R	3a–3R	IIIb	Not safe, low risk
Class 3B	3B	IV	Hazardous. Viewing of diffuse reflection is safe
Class 4	4	IV	Hazardous. Viewing of diffuse reflection is also hazardous. Fire risk

7.4. Radiation Exposure and Laser Goggle Selection

After the recognition of the laser hazards, protective safety design should be followed; calculation of radiation exposure, determination of the nominal hazard zone and selection of proper protective eyewear are basic starting points. Protective housings, curtains, interlocks, viewing scopes, viewing cards, beam stops and protective windows can be utilised depending on the exposure level [32,33].

7.4.1. *Radiation exposure basics*

An important quantity of measures related to laser damage to the eye is retinal irradiance (E). It is defined as power per unit area and usually expressed in W/cm^2 or mW/cm^2.

$$E \ (W/cm^2) = \frac{Power \ (W)}{Area \ (cm^2)}$$

The irradiance is much higher at the retina than at the cornea or lens because the laser light is focused onto a small area of the retina. The reduced size of the irradiated area indicates higher values of irradiance. Example 1 shows the calculation of the irradiance at the retina.

Example 1
Calculate the retinal irradiance when a laser of 1.0 mW average power enters the eye and is focused by the lens onto a small spot on the retina with a diameter of 10 μm.

Given: $P = 1$ mW, $D = 10$ μm

Solution: $A = \pi \left(\frac{D}{2}\right)^2 = \pi \left(1.0 \times 10^{-3} cm / 2\right)^2$

$$= 3.14 \times 10^{-6} / 4 = 7.85 \times 10^{-7} \ cm^2$$

$$E = \frac{P \ (W)}{A \ (cm^2)} = \frac{1.00 \times 10^{-3}}{7.85 \times 10^{-7}} = 1.27 \times 10^5 \ W/cm^2$$

When the eye is exposed to the laser, the reflex mechanism can prevent further exposure. Therefore, energy density is also an important quantity in the estimation of the eye damage;

$$Energy \ Exposure \ (J/cm^2) = E \ (W/cm^2) \cdot \Delta t \ (s)$$

where, Δt is the exposure time to the laser beam.

Example 2 shows the energy exposure when the exposure time is relatively short thanks to the reflex.

Example 2
Calculate the energy exposure when the exposure time is 0.25 s thanks to the reflex mechanism. The other conditions are the same with Example 1.

Given: $E = 1.27 \times 10^5 \ W/cm^2$, $\Delta t = 0.25 \ s$

Solution: $Energy \ Exposure = E \ (W/cm^2) \cdot \Delta t \ (s)$
$$= (1.27 \times 10^5) \times 0.25$$
$$= 3.18 \times 10^4 \ J/cm^2$$

7.4.2. *Maximum permissible exposure*

Maximum permissible exposure (MPE) is the level of light to which a person may be exposed without risk of injury. MPE is a function of many parameters, not just the intensity, but also the wavelength and the pulse duration. There are detailed sets of rules for calculating MPE for a given situation; such rules are occasionally revised according to new scientific findings. It is sometimes set to be 10% of the laser parameters for which there is a 50% probability of damage. The MPE should be measured at the cornea of the human eye or on the skin, for a given wavelength and exposure time. The calculation of the MPE follows the steps as listed below.

MPE calculation steps
1. Determine the radiation wavelength.
2. Determine maximum exposure level (intensity).
3. Consider the pulsing effect when the radiation is pulsed.
4. Calculate the MPE from charts.
5. Compare the MPE to the calculated exposure.

Because the eye and the skin have different absorption rates and damage mechanisms, the radiation wavelength must first be considered. For example, for the eye, UV-A and UV-B light cause accumulating damage, even at low powers, whereas IR-B and IR-C with wavelengths larger than 1.4 μm are absorbed by the lens and the cornea before reaching the retina; this means that the MPE for longer wavelengths (IR-B and IR-C) is higher than for visible and ultraviolet light. For the skin, MPEs are basically the same outside the retinal hazard region, from 0.4 to 1.4 μm, as shown in Fig. 7.14.

The maximum exposure level to the retina or the skin should be considered next, in the unit of W/cm^2; spatial distribution of the light needs to be taken into account during this consideration. Collimated laser beams cause serious injuries even at relatively low powers because the lens focuses the beam onto a small spot on the retina. For divergent radiation, such as the light from an LED with low spatial coherence, the light spreads over a large focal area on the retina; so the MPE is higher than for the collimated laser beams. In the MPE calculation, the worst-case scenario is assumed, in which the eye lens focuses the light into the smallest spot on the retina and the pupil is fully open. Although the MPE is specified as retinal irradiance (the power per unit surface), it is based on the power that can pass through a fully open pupil (0.39 cm^2) for visible and NIR wavelengths. This is relevant for laser beams that have a cross-section smaller than 0.39 cm^2.

If a laser is operated in pulsed mode, the peak power of the laser beam can be much higher than the average power depending on the pulse repetition rate and pulse duration. Therefore, the injury to the eye or to the skin is more serious when a pulsed laser beam is involved in the damage mechanism than a continuous-wave beam. The pulse repetition rate effect can be considered by,

$$MPE\ (Pulsed) = MPE(Continous) \cdot N^{-1/4}$$

where N is the number of pulses. The pulse duration effect can be further considered by the chart as shown in Figs. 7.14 and 7.15. They show the exposure-time-dependent MPEs, which is also important for long-term exposure cases. The IEC-60825-1 and ANSI Z136.1 standards include methods of calculating MPEs. For visible and NIR laser beams, sub-damage threshold effects, such as distraction, dazzle, glare and afterimages, also need to be considered.

Example 3
Calculate the maximum 'safe' average power when the eye is exposed to laser pulses with a 80 kHz repetition rate at a 1.0 μm centre wavelength for the exposure time of 1 ms.

Given: f_r = 80 kHz, λ = 1.0 μm, Δt = 1 ms

Solution:

$$
\begin{aligned}
N &= f_r\ (\text{Hz}) \cdot \Delta t(s) \\
&= 80{,}000\ (\text{pulses/s}) \times 0.001\ (s) \\
&= 80\ \text{pulses}
\end{aligned}
$$

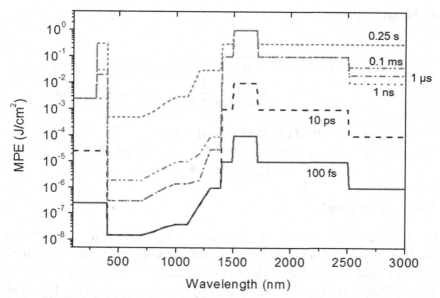

Fig. 7.14. Pulse-duration-dependent maximum permissible energy levels.

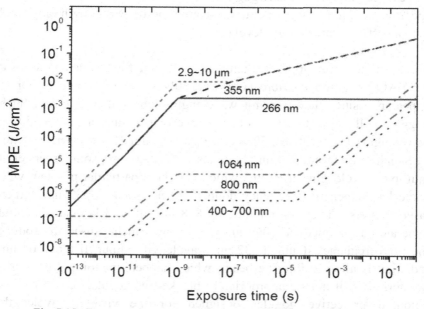

Fig. 7.15. Exposure-time-dependent maximum permissible energy levels.

In the MPE chart,

$$MPE(Continous) = \frac{P\ (W)}{A\ (cm^2)}$$
$$= \frac{1.00 \times 10^{-3}}{7.85 \times 10^{-7}} = 1.27 \times 10^5\ W/cm^2$$

$$MPE\ (Pulsed) = MPE(Continous) \cdot N^{-1/4}$$
$$= (1.27 \times 10^5) \cdot (80^{-1/4})W/cm^2$$
$$= 4.25 \times 10^4\ W/cm^2$$

7.4.3. *Laser goggles*

Protective eyewear such as laser goggles providing wavelength-selective filtering or power attenuation can protect our eyes from the reflected or scattered laser light, as well as from direct exposure to a hazardous laser beam. Therefore, the use of laser goggles when operating lasers of classes 3B and 4 is required in the workplace where the exposure to the eye exceeds the MPE. Laser goggles should be carefully selected for the specific laser, to block or attenuate the appropriate wavelength range and with a sufficient attenuation level (see Fig. 7.16).

For example, laser goggles attenuating 532 nm for frequency-doubled Nd:YAG or Yb-doped fibre lasers typically show an orange colour in appearance since they transmit wavelengths longer than 550 nm. Such goggles will be useless as protection eyewear against a laser emitting wavelengths longer than 550 nm, e.g. a 633 nm HeNe laser, 800 nm Ti:Sapphire laser and 980 nm diode laser. Furthermore, some lasers emit multiple wavelengths at a time, which may be a particular problem with some less expensive frequency-doubled lasers, such as 532 nm "green laser pointers". They are pumped by 808 nm infrared laser diodes, and generate an intermediate 1064 nm laser beam which is used to produce the final frequency-doubled 532 nm wavelength output. If the 1064 nm radiation is allowed into the beam, which happens in some green laser pointers, it will not be appropriately blocked by regular red or orange coloured protective goggles designed for the visible wavelength.

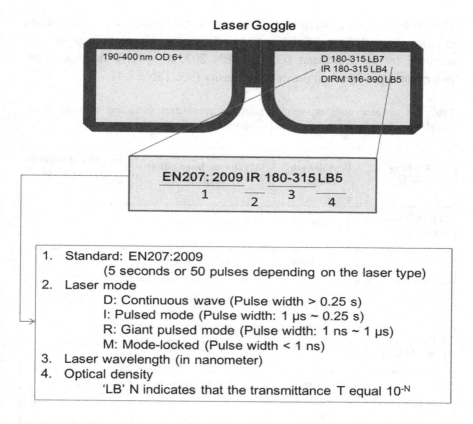

Fig. 7.16. Checking the protection capability of a laser goggle. Four important protection details are specified on the front surfaces of the laser goggles.

There are special laser goggles designed for covering dual wavelengths for frequency-doubled YAG and Yb-fibre lasers but they are rare and more expensive.

The other important factor that needs to be considered in selecting a laser goggle is the attenuation level. Laser goggles are rated for optical density (OD), which is the base-10 logarithm of the attenuation level by which the optical filter attenuates the incident beam power. For example, eyewear with OD 5 will reduce the beam power by a factor of 100,000. Furthermore, laser goggles must withstand a direct hit from the energetic laser beam without breakage.

The protective specifications (wavelengths and ODs) are printed on the goggles, near the top of the eyewear. Laser goggle manufacturers are required by the European standard, EN 207, to specify the maximum power rating rather than the optical density (see Table 7.4).

Table 7.4. Laser goggles: various selection parameters including working mode, wavelength, optical power density and minimum protection levels.

Working mode	Wavelength range	Maximum laser power density	Minimum protection level for given power*
D (continuous-wave laser)	180–315 nm	$1 \times 10^{n-3}$ W/m^2	$\log(P) + 3$
	315–1400 nm	$1 \times 10^{n+1}$ W/m^2	$\log(P) - 1$
	1400 nm –1000 µm	$1 \times 10^{n+3}$ W/m^2	$\log(P) - 3$
I, R (pulsed laser)	180–315 nm	$3 \times 10^{n+1}$ J/m^2	$\log(E/3) - 1$
	315–1400 nm	$5 \times 10^{n-3}$ J/m^2	$\log(E/5) + 3$
	1400 nm –1000 µm	$1 \times 10^{n+2}$ J/m^2	$\log(E) - 2$
M (ultrashort pulsed laser)	180–315 nm	$1 \times 10^{n+10}$ W/m^2	$\log(P) - 10$
	315–1400 nm	$1.5 \times 10^{n-4}$ J/m^2	$\log(E/1.5) + 4$
	1400 nm –1000 µm	$1 \times 10^{n+11}$ W/m^2	$\log(P) - 11$

* P in W/m^2 and E in J/m^2
Level numbers should be rounded upwards.
n=0 indicates that no safety goggles are necessary.

Example 4
Specify a pair of laser safety goggles to block a 633 nm HeNe laser beam with an average power of 1 mW and a 2 mm beam diameter.

Given: $P_{avg} = 1$ mW, $\lambda = 633$ nm, $D = 2$ mm

Solution:
$$E = \frac{P\ (W)}{A\ (m^2)} = \frac{1.0 \times 10^{-3}\ (W)}{\pi\ (2.0 \times 10^{-3}/2)^2\ (mm^2)}$$
$$= 3.18 \times 10^2\ W/mm^2$$

For continuous wavelength (D) of $\lambda = 633$ nm, the minimum protection level (OD) for the given power can be determined as

$$Optical\ Density\ (OD), n = \log(P) - 1$$
$$= \log(3.18 \times 10^2) - 1 = 1.5$$

which can be rounded up to 2. Therefore, the laser goggles' rating is

$$EN207\ D\ 633\ L2.$$

7.5. Precautions and Exemplary Hazardous Cases

Everyone who uses lasers should be aware of the possible risks so as to protect himself or herself from injury. This section shows a number of examples of technical precautions and hazardous cases.

7.5.1. *Technical precautions*

Examples of frequently used technical laser safety precautions are:

- ✓ Door level
  - ➤ Warning signs, warning lights and automatic door locks.
- ✓ Entry level
  - ➤ Protective goggles providing strong optical attenuation at laser(s)'s wavelengths.

✓ Laser level
  ➢ Encapsulation of laser systems by absorbing materials
  ➢ Key-operated power supply switches to prevent unauthorised use.
  ➢ Electronic interlocks that automatically switch-off lasers or block laser beams.
✓ Table level
  ➢ Protective housings, e.g. monitoring the presence of persons with measures such as light curtains, laser scanners or people counters.
  ➢ Beam stoppers (not only for main beams but also for parasitic reflections), preventing dangerous beams from leaving the optical setup.
✓ Alignment strategy
  ➢ For Class 3B or 4 lasers, trained personnel should handle their alignment.
  ➢ Laser experiments must be carried out on an optical table.
  ➢ Laser beams should travel only in the horizontal plane.
  ➢ All beams must be stopped at the edges of the table.
  ➢ Laser users should not put their eyes at the beam height.
  ➢ Never intentionally look directly into a laser.
  ➢ Do not stare at the light from any laser.
  ➢ Allow yourself to blink if the light is too bright.
  ➢ Never direct the beam towards other people.
  ➢ Low-power visible pilot beams should be used during the initial alignment.
  ➢ Beam path of dangerous and strong invisible laser radiation should be marked on the optical table.
  ➢ Watches, jewellery or other reflective material that might enter the optical plane should not be allowed in the laboratory.
  ➢ All non-optical objects that are close to the optical plane should have a matte finish in order to prevent specular reflections.
  ➢ High-intensity beams that can cause damage need to be guided through opaque tubes.

7.5.2. *Examples of hazardous situations*

Let us consider exemplary cases of hazardous situations as listed below. This list does not include the complete set of examples; but it is intended to improve user awareness of the multitude of possible hazards [34–37]:

- ✓ Electrical hazards
  - ➤ High-voltage power supplies to laser controller can be dangerous if workers can come into contact with the internal electronic unit or with a defective high-voltage cable.
- ✓ Chemical hazards
  - ➤ Some lasers require the handling of hazardous chemicals, e.g. carcinogenic dye solutions in dye lasers. Some of these solutions can penetrate the skin, and therefore need to be handled with special care.
- ✓ Optical hazards
  - ➤ Near-infrared laser beams are much more hazardous than visible light with the same power level because their radiation is focused on the retina just in the same way as visible light; but the blinking reflex of the human eye (i.e. normal closing of the eyelid quickly when the intensity is too high) is not active. Also, no warning is possible e.g. through weak stray light: nothing can be seen when a dangerous beam propagates in an unexpected direction.
  - ➤ Ultraviolet lasers endanger not only eyes but also the skin (see above).
  - ➤ Pulsed laser sources, e.g. Q-switched lasers or regenerative amplifiers, generate pulses with a peak power many orders of magnitude higher than even the average output power of a high-power laser. A single pulse from a hand-held miniature laser can totally destroy an eye.
  - ➤ In open laser setups, parasitic specular reflections (caused either by parts of the setup or by movable

metallic tools, watchbands, rings, etc., or by the residual reflectivity of anti-reflection coatings) may allow hazardous beams to leave the setup and potentially hit someone's eye.

➢ Optical fibres, e.g. transporting high optical powers between different rooms, may release dangerous radiation when being damaged. They, therefore, need to be specially protected and marked.

➢ High-power lasers (e.g. with powers in the kilowatt region) can damage not only the eye but also the skin within short exposure times, and can easily start a fire, e.g. when the beam hits materials such as wood or plastics; toxic fumes may also be generated.

✓ Human factors

➢ Remove your watch, jewellery and so on.

➢ Use adequate tools.

➢ Return tools after use.

➢ Remove scrap and spare parts to their designated places.

➢ Keep documents and desks in order.

➢ Deep working areas clean.

➢ Use personal protective equipment when specified.

➢ Reset controls after machine use.

➢ Keep records of maintenance and interruptions.

Problems

1. Explain why a laser beam is much more dangerous to our eyes and skin than lamp light. Can we use lamp lights instead of a laser beam in 3D printing and manufacturing? What is the reason?

2. How small a spot can we focus the light into, using a laser and using a thermal lamp? What is the fundamental limitation behind the limited focusing capability?

3. Explain the difference between photo-thermal and photo-chemical damages.

4. Describe how our eye reflex mechanism can contribute to preventing eye damage. Explain the details at different wavelength ranges.

5. Different spectral absorption at different parts in the eye provides different damage mechanisms. What are the dominant damaging wavelengths for the cornea, lens, vitreous and retina?

6. Select the appropriate laser classes for these examples.

   6.1. 0.2-mW laser diode

   6.2. Well enclosed 10 W Yb-fibre laser

   6.3. 1-mW HeNe laser used for optical interferometry

   6.4. 1 kW Yb-fiber laser for selective laser sintering

   6.5. 10 W UV laser for a stereolithography apparatus.

7. Explain the maximum permissible exposure (MPE) in the wavelength, pulse duration and exposure time domains. How can we select proper protective devices using MPE?

8. A laser technician wants to select a pair of goggles for a 532 nm and 10 ns pulse laser with an optical density higher than 7. Suggest proper goggles based on the EN207:2009 standard.

References

[1]   Hecht, J. (1986). *The laser guidebook.*

[2]   Yariv, A. (1976). *Introduction to optical electronics.*

[3]   Weber, M. J. (2000). *Handbook of lasers* (Vol. 18). CRC press.

[4]   Saleh, B. E., Teich, M. C., & Saleh, B. E. (1991). *Fundamentals of photonics* (Vol. 22). New York: Wiley.

[5]   Paschotta, R. (2008). *Encyclopedia of laser physics and technology.* Berlin: Wiley-vch.

[6]   Occupational Safety & Health Administration, U.S. department of labor, *Technical Manual on Laser Hazards*, http://www.osha.gov/dts/osta/otm/otm_iii/otm_iii_6.html

[7]   Berry, E. (2003). *Risk perception and safety issues*. Journal of biological physics, **29**(2), 263–267.

[8]   Laser Institute of America on laser safety, http://www.lia.org/subscriptions/safety_bulletins

[9]   Sliney, D. H. & Trokel, S.L. (1992). *Medical lasers and their safe use*. Springer Verlag ISBN 3540978569

[10]   Sliney D. H. & Wolbarsht M. L. (1980). *Safety with lasers and other optical sources*. Plenum Press 1980 ISBN 0306404346

[11]   Henderson, R. & Schulmeister, K. (2004). *Laser safety. IOP Publishing* ISBN 9780750308595

[12]   Hoss, R. J. and Lacy E. A. (1993). *Fiber optics*. Prentice-Hall PTR ISBN 0133212416

[13]   McKinlay, A. F., Harlen, F. & Whillock, M.J. (1998). *Hazards of optical radiation: a guide to sources, uses and safety*. Adam Hilger ISBN 0852742657 (Currently out of print)

[14]   Moseley, H. O. & Davies W. (2003). Biomedical laser safety. In: Webb CE and Jones JDC (eds.): Handbook of laser technology and applications. IOP Publishing 2003 ISBN 9780750309660

[15]   Moseley, H. (1988). *Non-ionising radiation*. Hilger.

[16]   BS EN 60825-1 *Safety of laser products, Part 1: Equipment classification and requirements*.

[17]   BS EN 60825-2 *Safety of laser products, Part 2: Safety of optical fibre communications systems*.

[18]   PD IEC/TR 60825-3 *Safety of laser products, Part 3: Guidance for laser displays and shows*.

[19]   PD IEC/TR 60825-14:2004 *Safety of laser products, Part 14: A user's guide*.

[20]   D CLC/TR 50448:2005 *Guide to levels of competence required in laser safety*.

[21]   International Electrotechnical Commission: IEC 62471:2006 *Photobiological safety of lamps and lamp systems*. IEC 2006.

[22]   BS EN 60825-4:2006 *Safety of laser products*. Laser guards.

[23]   PD IEC/TR 60825-8:2006 *Safety of laser products. Guidelines for the safe use of laser beams on humans*. BSI 2006.

[24]   BS EN ISO 11990:2003 *Optics and optical instruments. Lasers and laser-related equipment. Determination of laser resistance of tracheal tube shafts*. BSI 2003.

[25]   BS EN ISO 14408:2005 *Tracheal tubes designed for laser surgery. Requirements for marking and accompanying information*. BSI 2005. CRC 2006 ISBN 9780824723071

[26]   BS EN ISO 11810-1:2005 *Lasers and laser-related equipment. Test method and classification for the laser resistance of surgical drapes and/or patient protective covers. Primary ignition and penetration*. BSI 2005.

[27]   BS EN ISO 11810-2:2007 *Lasers and laser-related equipment. Test method and classification for the laser-resistance of surgical drapes and/or patient-protective covers. Secondary ignition*. BSI 2007.

[28]   Carruth JAS and McKenzie AL: *Medical lasers: science and clinical practice*. Adam Hilger 1986 ISBN 0852745605

[29]   BS EN 60601-1-1:2001 *Medical electrical equipment. General requirements for safety. Collateral standard. Safety requirements for medical electrical systems*. BSI 2001.

[30]   International Electrotechnical Commission (IEC), *Origin of international laser safety standards*: IEC 60825-1 ("Safety of laser products – Part 1: Equipment classification, requirements and user's guide") and IEC-60825-2 ("Safety of laser products – Part 2: Safety of optical fibre communication systems (OFCS)"), IEC, Geneva

[31]   American National Standards Institute (ANSI), http://www.ansi.org/, origin of the American Z-136 safety standard series, in particular the important Z-136.1

[32]   BS EN 207:1999 *Personal eye-protection. Filters and eye-protectors against laser radiation (laser eye-protectors)*. BSI 1999. MHRA DB2008(03) April 2008 60/81

[33]   BS EN 208: 2009 *Personal eye-protection – Eye-protectors for adjustment work on lasers and laser systems (laser adjustment eye-protectors)*.

[34]   Medicines and Healthcare products Regulatory Agency. DB 2006(05) *Managing Medical Devices*. MHRA 2006. www.mhra.gov.uk

[35]   Sulieman, M., Rees, J. S., & Addy, M. (2006), *Surface and pulp chamber temperature rises during tooth bleaching using a diode laser.* British Dental Journal, **200**(11) p. 631–634.

[36]   Wheeland, R.G. (1995). *Cosmetic use of lasers. Dermatologic Clinics* **13**(2), 447-59.

[37]   Savastru, D., Miclos, S. et al (2006). *Microsurgical Nd:YAG laser used in ophthalmology*. Romanian Journal of Physics, **51**(7–8), 833–837.

Chapter 8

FUTURE PROSPECTS

The global 3D printing market is expected to grow at a compound annual growth rate (CAGR) of 32% to €20 billion by 2020. In this ever-increasing market, the portion of laser-based 3D printing in the overall 3D printing market is increasing as well. This is because, metal-based 3D printing is growing at higher rates than polymer-based 3D printing. From 2010, the major applications of 3D printing are moving from product design (i.e. prototyping and customisation) to the actual production of end products. For 3D printing of functional parts and units, metals are better than polymers in material characteristics. A high-power laser beam (or electron-beam) is the most appropriate energy source for 3D printing of metals (and ceramics) because of their high melting temperature. Here, in this chapter, we will discuss the trends of laser-based 3D printing technology and the applications in which 3D printing could find its role in near future.

8.1. Introduction

3D printing technology has evolved over the past 30 years [1–3]. As opposed to conventional manufacturing, 3D printing provides new abilities to produce parts with complex configurations and accommodate design changes with high flexibility; to meet rapidly changing industrial demands, without incurring additional tooling costs or time delays. Typically, products can be produced within a few hours to a week, and additive manufacturing does not require high volume production to break even. As such, 3D printing is understood as the ideal technology for addressing dynamic technological trends and industrial demands. Laser is one of the best energy sources in 3D printing for maximising its potential. Some examples showing the laser's capabilities are listed as follows. First, for printing very fine structures (e.g. micro-channels or nozzles in turbine engine parts for the aerospace or defence industries or hydrophobic/hydrophilic surfaces for the marine and offshore industry), high resolution 3D printing is a prerequisite. With the aid of the high coherence of the laser beam, one can make the fine structure with sub-micrometre dimensional resolution. Second, for higher 3D printing productivity in SLS or SLM, the powders should be melted within a short time. However, it is not an easy task for metallic and ceramic powders due to their high melting temperatures and high-level heat capacities. When considering a 100 W continuous-wave laser focused on a metallic target material with 10% absorption, the material temperature can exceed 1,500°C within an exposure time of 0.001 s. With pulsed lasers, this temperature increase can be even faster and more efficient. Third, 3D printing of multiple materials is getting higher attention. Here, 'multi-material' does not mean the same material with different colours but a combination of different materials. Today's micro-electronics products are good examples; they are composed of an electrically insulating polymer, electrically conducting metal electrodes and a ceramic structure for micro capacitors. With the aid of a special laser system, multi-material 3D printing is expected to be realised in the near future. In summary, lasers are one of the best energy sources for 3D printing for providing high-resolution, high throughput and multi-material printing capabilities at the time of demand. In this chapter, technological trends in the next-generation of 3D printing will be discussed, as well as further discussion on future applications of 3D printing.

8.2. Trends in Next-Generation 3D Printing

The next-generation 3D printing market is expected to require a high-precision, high-strength, fully-functional, multi-scale and multi-material 3D printing product at the time of demand [1,4,5]. To meet these requirements, there are critical challenges to overcome in 3D printing: low resolution, slow speed and limited range of 3D printable materials. Laser-based 3D printing is the most attractive solution for highly-flexible next-generation 3D printing, but the performance has been limited by available lasers in the market. As covered earlier, the representative laser systems are CO_2 lasers, Nd:YAG lasers and Excimer lasers. In the near future, new highly-efficient multi-colour laser systems will become available over broader spectral bandwidths, which will improve the performance of 3D printed products and facilitate their wider use. Four trends for future laser-based 3D printing are listed as follows (see Fig. 8.1):

- High resolution
- High productivity
- Various materials (and combinations of different materials)
- Print on demand.

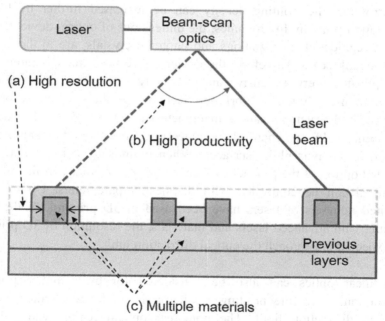

Fig. 8.1 Conceptual diagram for the next-generation of laser-based 3D printers.

We will discuss the potential and possibilities of laser-based 3D printing with these trends one by one as follows.

8.2.1. *High resolution*

The resolution of laser-based 3D printing is determined by a focused spot size, powder size, heat affected zone (HAZ) and/or beam scanning speed in the lateral X-Y domain. When the other technological parameters are optimised, the fundamental and final resolution limit comes from the optical diffraction limit. If the laser beam is in a well-defined spatial mode (e.g. TEM00 mode), the best focal spot size is about half the wavelength (by optical diffraction limit). This high printing resolution level is beneficial for manufacturing various interesting structures for electronics, photonics and mechanical engineering. In electronics, supercapacitors for micro-batteries require highly dense metallic electrodes for a wider surface area for an efficient interaction with electrolytes; the printing density can be increased further by higher resolution printing. In photonics, the dimensions of optical devices, such as waveguides, Bragg gratings and photonic crystals, are in the optical wavelength scale. Therefore, the devices have been manufactured by a complicated, very expensive and relatively slow lithography process. With high-resolution 3D printing technology, these complex optical devices and structures can be manufactured with a high-level of design flexibility. In mechanical engineering, one good example is hydrophobic/hydrophilic surfaces. When optimised microstructures are printed on top of the product surfaces, the hydrophobicity of the original material can be controlled with the design parameters. Up to now, limited numbers of lasers have been used in 3D printing. In the near future, with advanced lasers and materials, the resolution of 3D printing is expected to approach the optical diffraction limit.

Nonlinear optics can also be considered for even higher printing resolution. Here, the printing resolution is no longer limited by the optical diffraction limit. The nonlinear optical regime can only be reached when using ultra-short (nanosecond, picosecond and femtosecond) pulse lasers. As covered earlier, two-photon polymerisation or multi-photon polymerisation are based on nonlinear optics, which provides a sub-100 nm printed feature size. Other than multi-photon polymerisation, the 3D printing of nano/micro metallic structures has been recently reported to use a femtosecond laser as the light source; for example, using the photo-aggregation effect in a metal-ion solution and the efficient photo-reduction of graphene oxide (GO) to reduced graphene oxide (rGO). Therefore, ultra-short pulse lasers will be widely introduced to the field of 3D printing.

8.2.2. High productivity

The laser beam is used as the energy source in 3D printing for changing the phase of the target material, e.g. from liquid to solid in SLA; from powder to sintered/connected powder in SLS; and from powder to bulk in SLM. In all the laser-based 3D printing processes, higher energy transfer to target materials guarantees higher productivity. Basically, the laser is a good energy source especially for the 3D printing of metals. Considering that the melting temperatures of metals are about thousands of degree Celsius (e.g. melting temperatures: Aluminium: 659°C, Gold: 1063°C, Copper: 1083°C, Steel: 1371°C, Nickel 1452°C, wrought iron: 1482°C and Tungsten: 3399°C) and that the heat capacities of metals are generally high, melting the metals for 3D printing is not an easy task for conventional energy sources; lasers and electron-beams are the possible solutions for the purpose. Among them, lasers are more attractive in industries because they do not need a vacuum environment. The average power of the lasers in use in the field of 3D printing is getting higher, to achieve a higher productivity. Recently, high-power fibre lasers with a few kW average powers have been adopted.

When a laser beam is delivered to a target material, there is photon–matter interaction; some part of the beam is reflected, another part of the beam is transmitted and the other part of the beam is absorbed. Thus, for a higher energy transfer rate, the material should have a high absorption coefficient at the laser wavelength. Therefore, lasers with different wavelengths will be newly designed, developed and introduced to 3D printing. As an example, polycarbonates absorb 90% of the incident light at 1750 nm; this novel wavelength laser can support a spatial resolution that is 10 times higher than the traditional 10.6 μm wavelength based on the optical diffraction-limit with a minor absorption loss of about 10%.

Generation of a strong ultraviolet (UV) light by a nonlinear optical wavelength conversion is also expected to open a new possibility in 3D printing and nonlinear optical absorption can be utilised for the 3D printing of transparent nano/micro structures in addition to a linear optical absorption. When high-power ultra-short laser pulses are focused on the transparent material, nonlinear optical absorption will start to work.

The light absorptivity of 3D printing materials in SLS and SLM can be also increased by optimising the materials' properties. By changing the size, shape and coating properties of metallic or ceramic powders, material absorptivity can be further improved by multiple scattering, plasmonic resonance and nonlinear absorption effects inside the powder pool.

High energy concentration is required for an efficient phase change of the material as well. The absorbed photon energy in the material should be kept within a small volume without unexpected dissipation for efficient material phase change; this is enabled when using ultra-short pulse lasers. When ultra-short laser pulses are focused on the material, the absorbed heat cannot be dissipated out to the surrounding volume because the pulse duration is even shorter than the time required for any heat transfer through conduction. Thus, ultra-short pulse lasers are expected to be adapted for higher productivity 3D printing.

8.2.3. *Various materials and combination of multiple materials*

3D printing of multiple materials is expected to attract higher attention. Here, multi-material does not mean 'the same material in different colours' but 'a combination of different materials'. Today's micro-electronics products show good examples; micro-electronics typically contain an electrically insulating polymer layer, electrically conducting electrodes and a ceramic structure for micro-capacitors. Micro-electronics, including semiconductors, flat panel displays, flexible displays, printed circuit boards, solar cells and fuel cells share similar material compositions. Due to their different material properties (e.g.

absorption wavelength range, melting temperature and so on), a laser-based 3D printer designed for a single target material cannot print multi-material structures. For example, 3D printers for polymers with a 10.6 µm CO_2 laser cannot print metallic or ceramic parts due to low material absorption and relatively high melting temperatures. By using high-power UV lasers (because most materials strongly absorb UV), using a combination of several lasers in different wavelengths, or using nonlinear absorption with the aid of ultra-short pulse lasers, multi-material 3D printing is expected to be realised in the near future (see Fig. 8.2).

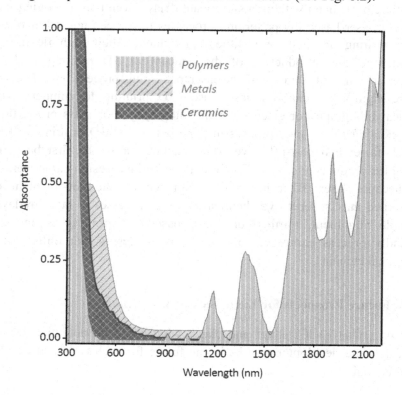

Fig. 8.2. Material absorption spectra of polymers, metals and ceramics.

The combination of multiple materials, such as a metals and ceramics, could be an ideal solution for high-strength and bio-compatible medical implants. The combination of different metals, such as a tungsten carbide exterior with a heat-dissipating copper inner layer, will also enable high-productivity injection moulding with a longer tool lifetime, as well as high-performance automobile and aerospace units. These combinations could be realised by using lasers as the energy source.

8.2.4. *Print on demand*

3D printing is basically a very flexible manufacturing technology and so is laser-based 3D printing. Therefore, one can print the target product at the time of demand without a significant delay. Even before moving into the 3D, laser-based 2D patterning processes have been termed as direct laser writing or direct laser lithography due to their high flexibility. Thanks to the introduction of the concept of 3D printing in laser patterning, the advantages of the laser patterning process have become more significant, enabling laser-based 3D printing to compete with optical lithography or electron-beam lithography. For mass production, the existing lithography process may be better, but 3D printing will be much faster and cost-effective when manufacturing the first batch of complex products or devices for finding the optimal design parameters. It is because laser 3D printing does not require the development of patterned masks, repetitive chemical etching processes or layer-by-layer sample alignment. Furthermore, laser-based 3D printing is an eco-friendly process compared to lithography processes that utilise toxic chemical etchants.

8.3. Future Prospects for Each Technical Chapter

Let us revisit each chapter and check what could be potential design, research and development tasks in the future for each technical chapter of this book.

8.3.1. *Lasers and basic optics*

- From ray optics to wave, electromagnetic and quantum optics
  There are four competing theories of light; ray optics, wave optics, electromagnetic optics and quantum optics. They are

in a relationship as follows: Ray Optics ⊂ Wave Optics ⊂ Electromagnetic Optics ⊂ Quantum Optics. To date, 3D printing focused mainly on ray optics. The inclusion of wave optics enables nano/micropatterning by using optical interference, as covered in Chapter 6 of this book. In electromagnetic optics, polarisation can be introduced for controlled light-matter interaction. If someone wants to design, develop or optimise lasers, quantum optics should be covered. An in-depth study of fundamentals of optics and photonics will open new possibilities.

- From linear to nonlinear optics
  By introducing short pulse lasers into 3D printing, nonlinear optics can be utilised. Nonlinear optics provides nonlinear absorption in transparent materials, which enables one to attain a higher wavelength conversion efficiency (in second-, third-harmonic, high harmonic generation or supercontinuum generation). As a result, the laser's wavelength and its absorption in the material can be controlled using nonlinear optics.

- Not just an energy source but as the source with high coherence
  Using the high temporal and spatial coherence of the light, one can make optical interference among several split beams, which can synthesise ultra-short pulses and focus the laser beam onto a small spot, less than 1 μm.

8.3.2. *Materials in laser-based 3D printing*

- Application-oriented functional materials
  Novel application-oriented materials need to be developed for the laser-based 3D printing of micro-electronics, printed circuit boards and planar optical devices.

- Multi-material composition
  The combination of different materials will be widespread for the 3D printing of functional devices.

● Interdisciplinary studies

For next-generation laser-based 3D printing, interdisciplinary studies among mechanical engineers, material engineers, physicists and biologists are expected to be essential.

● 3D printing of bio-materials
There is a huge research trend moving towards 3D printing bio materials for fundamental studies to understand tissue/cell characteristics and the manufacturing of functional human tissues for medical applications.

● 3D printing of carbon materials
Graphene and carbon nanotubes are receiving great attention for their potential applications in electronics, photonics, biomedical and energy storage devices, sensors and other cutting-edge technological fields, mainly because of their fascinating properties; such as extremely high electron mobility, good thermal conductivity and high elasticity.

● Metamaterials: synthetic optical absorption and transmission
By printing periodic metallic structures, a synthetic optical structure with the designed optical characteristics can be manufactured. Optical absorption, transmission and polarisation degree can be designed and controlled therein.

8.3.3. *Laser-assisted manufacturing*

● Cooperation with traditional manufacturing technologies
Laser-based 3D printing can cooperate with traditional subtractive or formative laser machining processes. This is nowadays called hybrid manufacturing.

● Adoption of existing concepts into 3D printing
Traditional laser cladding and laser engineered net shaping (LENS) have many things in common. Like this, existing well-established concepts in traditional laser manufacturing can be applied to 3D printing.

8.3.4. *Laser-based 3D printing*

- Laser-based hybrid manufacturing
  Different combinations of subtractive, formative and additive manufacturing processes can be tested for better printing performances.

- Introduction of high-power short-wavelength pulsed lasers
  Lasers having UV emission wavelengths can commonly be used for 3D printing of polymers, metals and ceramics. However, because there are limited numbers of gain materials in the UV wavelength, frequency-tripled high-power NIR lasers are widely used. Because wavelength conversion from 1060 nm (NIR) to 353 nm (UV) is a nonlinear optical process, the conversion efficiency increases in proportion to the input laser's peak power. Therefore, pulsed lasers are preferred for generating UV wavelengths as they usually provide a higher peak power than continuous-wave lasers. Wavelength conversion is made by focusing the laser beam into a nonlinear optical crystal, such as KTP, LBO or BBO.

- Nano/micro-printing based on multi-photon-absorption (MPA)
  With the aid of the high peak power of the femtosecond laser, two-photon polymerisation (TPP) can be realised. If the process is TPP, the second harmonic of the laser wavelength should lie in the absorption wavelength of photopolymers.

- Higher-power rare-earth-doped fibre lasers
  Lasers that have long been used in traditional laser machining, e.g. lamp-pumped Nd:YAG lasers, will be changed to diode-pumped high-power rare-earth-doped fibre lasers due to their high wall-plug efficiency, high system stability and easier maintenance.

8.3.5. *Advanced 3D manufacturing*

- Generalised patterns by applying phase engineering
  Two-, three- and four-beam interference already provides well-defined 3D pattern structures. In the future, generalised patterns are expected to be printed with the aid of phase engineering technologies.

8.4. Future Applications of Laser-based 3D Printing

In addition to traditional and newly opened 3D printing markets, such as the aerospace, automobile, defense and biomedical industries, laser-based 3D printing is expected to support the manufacturing of a broad range of future electronic products, such as printed micro-electronics, printed circuit boards, IoT devices, planner optical devices, solar cells and fuel cells (see Fig. 8.3). These applications require high resolution, high productivity, high functionality and/or multiple materials to 3D printing.

Fig. 8.3. Laser-based 3D printing of micro-electronics.

Micro-electronic products are commonplace in our daily lives, such as smartphones, smart watches, smart cars, flat panel displays, computers and various IoT devices. Inside these products, semiconductor chips and printed circuit boards exist. Semiconductor memory and non-memory chips are the core parts of these electronic devices which are costly and sensitive to environmental changes. So they must be packaged well to be protected from unexpected environmental influences. After packaging, semiconductor chips are connected to larger-sized printed circuit boards using metallic wires or 2D solid metal bump arrays. Laser-based 3D printing could help with the concept of 3D packaging, wiring or bonding.

Semiconductor chips are nowadays moving to stacked structures for a higher integration ratio; therefore, electrical connections among the layers become an important issue. 3D printing can be applied to this electrical connection among these stacked multiple layers (which requires the printing resolution of sub-μm to tens of μm) or for the interconnection between the packaged chip and the printed circuit board (which requires the printing resolution of tens of μm to several mm) [6–8].

Laser-based 3D printing can be utilised for the direct manufacturing of active and passive circuit components including resistors, inductors, capacitors, conformal coatings, transistors, sensors, micro-antennas and micro-batteries on printed circuit boards. Compared to existing circuit components manufactured by the screen-printing process, 3D-printed units can provide higher pattern resolution, an easier alignment process, design flexibility and new patterning possibilities on flexible substrates. As an example, thin film transistors for flexible displays, RF IDs and IoTs can be simply manufactured using laser-based multi-material 3D printing [9–10].

Our surroundings are getting smarter to better our lives thanks to smart devices. The Internet of Things (IoT) can be one good future application of laser-based 3D printing [11–13]. IoT devices could be in different forms but should be able to sense, collect, analyse and communicate data in common. This requires highly flexible manufacturing processes which can be satisfied with laser-based 3D printing. Important physical device components of IoT, such as sensors, micro-batteries, antennas and their interconnection parts can be 3D printed on general target substrates, even on curved or flexible ones.

3D printing technologies have already started to be applied to solar cell manufacturing [14–17]. Using laser as the energy source, finer features (e.g. collector electrodes and bus-bar lines) can be printed onto planar and non-planar substrates at ambient temperature without the need for masks or stencils. The narrower, high integrity collector lines will enable higher conductivity and a lower shadowing effect, thereby increasing the cell efficiency of photovoltaics. Fuel cells share similar structures with solar cells. Solid oxide fuel cells also get benefits from laser-based 3D printing by making fine or porous internal structures to provide a larger surface area interacting with oxygen; or by having better mechanical or thermal characteristics based on arbitrary-designed complex internal structures. Furthermore, laser-based 3D printing can print multiple materials in a mixed form to create smooth transitions between the material layers of a solid oxide fuel cell. This can provide a larger effective functional zone and better mechanical stability at the interfaces.

Next-generation optical circuits can also get benefits from laser-based 3D printing [18,19]. The simplest optical structure is an optical waveguide that can be 3D-printed by using a combination of two different polymers with different refractive indices. There have been a number of interesting demonstrations of optical waveguides since early 2000. Because optical structures require micrometre scale printing resolution, two-photon polymerisation, multi-photon polymerisation or interference lithography could be the base technologies. Based on the technologies, optical Bragg gratings, optical couplers, splitter, combiners, dividers and other passive optical devices can be manufactured.

Plasmonics is a new research field of photonics in polymer/metal interfaces. By applying multi-material printing capability of laser-based 3D printing, active photonic devices can be manufactured [20–26]. 3D printing of new combinations of polymer/metal structures will enable electrically controlled plasmonic optical modulators, intensity controllers, polarisation rotators, photo-detectors and so on (see Fig. 8.4).

8.5. Summary

In summary, lasers are one of the best energy sources for 3D printing and will definitely provide a high-resolution, high-throughput and multi-material printing capability at the time of demand to wider applications in very near future.

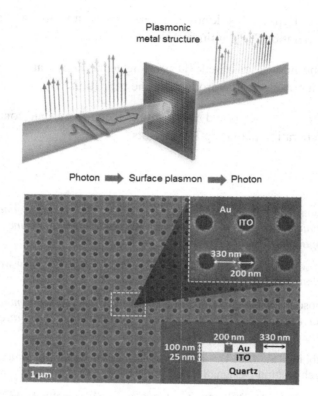

Fig. 8.4. Plasmonics as a laser-based 3D printing application.

Problems

1. List four future trends in next-generation laser-based 3D printing. Explain what benefits can lasers provide in these trends?

2. Explain what determines the resolution in laser-based 3D printing. How can one overcome the resolution limit of traditional laser-based 3D printing?

3. Describe how laser-based 3D printing provides better productivity in SLA and SLM.

4. List three representative lasers in traditional 3D printing markets and their centre wavelength. Which materials can be printed using those

lasers? Discuss what kind of lasers are promising candidates for next-generation 3D printing.

5. Describe how laser-based 3D printing technology can contribute to smart micro-electronic products in the near future.

6. Explain how laser-based 3D printing technology can contribute to next-generation planar optical circuits.

References

[1]   Chua, C. K., & Leong, K. F. (2015). *3D printing and additive manufacturing: principles and applications: principles and applications*. World Scientific, Singapore.

[2]   Lipson, H., & Kurman, M. (2013). *Fabricated: The new world of 3D printing*. John Wiley & Sons.

[3]   Gibson, I., Rosen, D., & Stucker, B. (2014). *Additive manufacturing technologies: 3D printing, rapid prototyping, and direct digital manufacturing*. Springer.

[4]   Tibbits, S. (2014). *4D Printing: Multi-Material Shape Change*. Architectural Design, **84**(1), 116–121.

[5]   Campbell, T., Williams, C., Ivanova, O., & Garrett, B. (2011). Could 3D printing change the world. *Technologies, Potential, and Implications of Additive Manufacturing*, Atlantic Council, Washington, DC.

[6]   Ko, S. H., Pan, H., Grigoropoulos, C. P., Luscombe, C. K., Fréchet, J. M., & Poulikakos, D. (2007). *All-inkjet-printed flexible electronics fabrication on a polymer substrate by low-temperature high-resolution selective laser sintering of metal nanoparticles*. Nanotechnology, **18**(34), 345202.

[7]   Muth, J. T., Vogt, D. M., Truby, R. L., Mengüç, Y., Kolesky, D. B., Wood, R. J., & Lewis, J. A. (2014). *Embedded 3D printing of strain sensors within highly stretchable elastomers*. Advanced Materials, **26**(36), 6307–6312.

[8]   Macdonald, E., Salas, R., Espalin, D., Perez, M., Aguilera, E., Muse, D., & Wicker, R. B. (2014). *3D printing for the rapid prototyping of structural electronics*. IEEE Access, **2**, 234–242.

[9]   Kawahara, Y., Hodges, S., Cook, B. S., Zhang, C., & Abowd, G. D. (2013, September). Instant inkjet circuits: lab-based inkjet printing to support rapid prototyping of UbiComp devices. *In Proceedings of the 2013 ACM international joint conference on Pervasive and ubiquitous computing* (pp. 363–372). ACM.

[10]   Siegel, A. C., Phillips, S. T., Dickey, M. D., Lu, N., Suo, Z., & Whitesides, G. M. (2010). *Foldable printed circuit boards on paper substrates.* Advanced Functional Materials, **20**(1), 28–35.

[11]   Gubbi, J., Buyya, R., Marusic, S., & Palaniswami, M. (2013). *Internet of Things (IoT): A vision, architectural elements, and future directions.* Future Generation Computer Systems, **29**(7), 1645–1660.

[12]   Vermesan, O., Friess, P., Guillemin, P., Gusmeroli, S., Sundmaeker, H., Bassi, A., ... & Doody, P. (2011). *Internet of things strategic research roadmap.*

[13]   O. Vermesan, P. Friess, P. Guillemin, S. Gusmeroli, H. Sundmaeker, A. Bassi, et al., *Internet of Things: Global Technological and Societal Trends,* **1**, 9–52.

[14]   Kildishev, A. V., Boltasseva, A., & Shalaev, V. M. (2013). *Planar photonics with metasurfaces.* Science, **339**(6125), 1232009.

[15]   O'regan, B., & Grfitzeli, M. (1991). *A low-cost, high-efficiency solar cell based on dye-sensitized colloidal TiO$_2$ films.* Nature, **353**(6346), 737–740.

[16]   Kim, H. S., Lee, C. R., Im, J. H., Lee, K. B., Moehl, T., Marchioro, A., ... & Grätzel, M. (2012). *Lead iodide perovskite sensitized all-solid-state submicron thin film mesoscopic solar cell with efficiency exceeding 9%.* Scientific Reports, **2**, 591.

[17]   You, J., Dou, L., Yoshimura, K., Kato, T., Ohya, K., Moriarty, T., ... & Yang, Y. (2013). *A polymer tandem solar cell with 10.6% power conversion efficiency.* Nature Communications, **4**, 1446.

[18]   Snyder, A. W., & Love, J. (2012). *Optical waveguide theory.* Springer Science & Business Media.

[19]   Chuang, S. L., & Chuang, S. L. (1995). *Physics of optoelectronic devices.*

[20]   Maier, S. A. (2007). *Plasmonics: fundamentals and applications.* Springer Science & Business Media.

[21]   Atwater, H. A., & Polman, A. (2010). *Plasmonics for improved photovoltaic devices.* Nature Materials, **9**(3), 205–213.

[22]   Schuller, J. A., Barnard, E. S., Cai, W., Jun, Y. C., White, J. S., & Brongersma, M. L. (2010). *Plasmonics for extreme light concentration and manipulation.* Nature Materials, **9**(3), 193-204.

[23]   Maier, S. A., Brongersma, M. L., Kik, P. G., Meltzer, S., Requicha, A. A., & Atwater, H. A. (2001). *Plasmonics—a route to nanoscale optical devices.* Advanced Materials, **13**(19), 1501–1505.

[24]   Ozbay, E. (2006). *Plasmonics: merging photonics and electronics at nanoscale dimensions.* Science, 311(5758), 189–193.

[25]   Park, I. Y., Kim, S., Choi, J., Lee, D. H., Kim, Y. J., Kling, M. F., ... & Kim, S. W. (2011). *Plasmonic generation of ultrashort extreme-ultraviolet light pulses.* Nature Photonics, **5**(11), 677–681.

[26]   Geng, X. T., Chun, B. J., Seo, J. H., Seo, K., Yoon, H., Kim, D. E., Kim, Y.-J. & Kim, S. (2016). *Frequency comb transferred by surface plasmon resonance.* Nature Communications, **7**, 10685.

Printed in the United States
By Bookmasters